Scatter Search
Methodology and Implementations in C

OPERATIONS RESEARCH/COMPUTER SCIENCE INTERFACES SERIES

Series Editors

Professor Ramesh Sharda
Oklahoma State University

Prof. Dr. Stefan Voß
Universität Hamburg

Other published titles in the series:

Greenberg, Harvey J. / *A Computer-Assisted Analysis System for Mathematical Programming Models and Solutions: A User's Guide for ANALYZE*
Greenberg, Harvey J. / *Modeling by Object-Driven Linear Elemental Relations: A Users Guide for MODLER*
Brown, Donald/Scherer, William T. / *Intelligent Scheduling Systems*
Nash, Stephen G./Sofer, Ariela / *The Impact of Emerging Technologies on Computer Science & Operations Research*
Barth, Peter / *Logic-Based 0-1 Constraint Programming*
Jones, Christopher V. / *Visualization and Optimization*
Barr, Richard S./ Helgason, Richard V./ Kennington, Jeffery L. / *Interfaces in Computer Science & Operations Research: Advances in Metaheuristics, Optimization, & Stochastic Modeling Technologies*
Ellacott, Stephen W./ Mason, John C./ Anderson, Iain J. / *Mathematics of Neural Networks: Models, Algorithms & Applications*
Woodruff, David L. / *Advances in Computational & Stochastic Optimization, Logic Programming, and Heuristic Search*
Klein, Robert / *Scheduling of Resource-Constrained Projects*
Bierwirth, Christian / *Adaptive Search and the Management of Logistics Systems*
Laguna, Manuel / González-Velarde, José Luis / *Computing Tools for Modeling, Optimization and Simulation*
Stilman, Boris / *Linguistic Geometry: From Search to Construction*
Sakawa, Masatoshi / *Genetic Algorithms and Fuzzy Multiobjective Optimization*
Ribeiro, Celso C./ Hansen, Pierre / *Essays and Surveys in Metaheuristics*
Holsapple, Clyde/ Jacob, Varghese / Rao, H. R. / *BUSINESS MODELLING: Multidisciplinary Approaches — Economics, Operational and Information Systems Perspectives*
Sleezer, Catherine M./ Wentling, Tim L./ Cude, Roger L. / *HUMAN RESOURCE DEVELOPMENT AND INFORMATION TECHNOLOGY: Making Global Connections*
Voß, Stefan, Woodruff, David / *Optimization Software Class Libraries*
Upadhyaya et al/ *MOBILE COMPUTING: Implementing Pervasive Information and Communications Technologies*
Reeves, Colin & Rowe, Jonathan/ *GENETIC ALGORITHMS—Principles and Perspectives: A Guide to GA Theory*
Bhargava, Hemant K. & Ye, Nong / *COMPUTATIONAL MODELING AND PROBLEM SOLVING IN THE NETWORKED WORLD: Interfaces in Computer Science & Operations Research*
Woodruff, David L./ *NETWORK INTERDICTION AND STOCHASTIC INTEGER PROGRAMMING*
Anandalingam, G. & Raghavan, S./ *TELECOMMUNICATIONS NETWORK DESIGN AND MANAGEMENT*

Scatter Search

Methodology and Implementations in C

Manuel Laguna
University of Colorado

Rafael Martí
University of Valencia

Kluwer Academic Publishers
Boston • Dordrecht • London

Distributors for North, Central and South America:
Kluwer Academic Publishers
101 Philip Drive
Assinippi Park
Norwell, Massachusetts 02061 USA
Telephone (781) 871-6600
Fax (781) 871-9045
E-Mail: kluwer@wkap.com

Distributors for all other countries:
Kluwer Academic Publishers Group
Post Office Box 322
3300 AH Dordrecht, THE NETHERLANDS
Telephone 31 786 576 000
Fax 31 786 576 254
E-mail: services@wkap.nl

 Electronic Services <http://www.wkap.nl>

Library of Congress Cataloging-in-Publication Data

A C.I.P. Catalogue record for this book is available from the Library of Congress.

Laguna, Manuel and Marti, Rafael/ SCATTER SEARCH: Methodology and
   Implementations in C

ISBN 1-4020-7376-3

Printed on acid-free paper.

Printed in the United States of America.

To Sofía and Fabián, my limitless source of happiness and inspiration

M. L.

A Mila, por la paz, y sobre todo, por la guerra que me da

R. M.

Contents

Foreword

The book Scatter Search by Manuel Laguna and Rafael Martí represents a long-awaited "missing link" in the literature of evolutionary methods. Scatter Search (SS)—together with its generalized form called Path Relinking—constitutes the only evolutionary approach that embraces a collection of principles from Tabu Search (TS), an approach popularly regarded to be divorced from evolutionary procedures. The TS perspective, which is responsible for introducing adaptive memory strategies into the metaheuristic literature (at purposeful level beyond simple inheritance mechanisms), may at first seem to be at odds with population-based approaches. Yet this perspective equips SS with a remarkably effective foundation for solving a wide range of practical problems. The successes documented by Scatter Search come not so much from the adoption of adaptive memory in the range of ways proposed in Tabu Search (except where, as often happens, SS is advantageously coupled with TS), but from the use of strategic ideas initially proposed for exploiting adaptive memory, which blend harmoniously with the structure of Scatter Search.

From a historical perspective, the dedicated use of heuristic strategies both to guide the process of combining solutions and to enhance the quality of offspring has been heralded as a key innovation in evolutionary methods, giving rise to what are sometimes called "hybrid" (or "memetic") evolutionary procedures. The underlying processes have been introduced into the mainstream of evolutionary methods (such as genetic algorithms, for example) by a series of gradual steps beginning in the late 1980s. Yet this theme is an integral part of the SS methodology proposed a decade earlier, and the form and scope of such heuristic strategies embedded in SS continue to set it apart. Although there are points in common between SS and other evolutionary approaches, principally as a result of changes that have brought other approaches closer to SS in recent years, there remain differences that have an important impact on practical outcomes.

Reflecting this impact, a hallmark of the present book is its focus on practical problem solving. Laguna and Martí give the reader the tools to create an SS implementation for problems from a wide range of settings. Although theoretical problems (such as abstract problems in graph theory) are included, beyond a doubt the practical realm has a predominant role in this book.

In pursuing this vision, the authors have provided an extensive body of illustrative computer code to show how important strategic elements can be implemented. The book is organized to provide a "hands-on" experience for the reader. Because SS is less well-known than most other evolutionary methods, this feature can prove valuable even for relatively experienced metaheuristic researchers who seek to understand and apply the method.

At the same time, Laguna and Martí go beyond the basics of SS. Following a series of helpful tutorial chapters (2, 3 and 4) they provide a particularly lucid treatment of advanced strategies in Chapter 5. Both researchers and practitioners will encounter a good deal of useful information in this transition from basic to advanced considerations. The exposition is designed to equip the reader to duplicate the experiences of those who have achieved highly effective past applications, and to take these experiences to another level, producing new implementations that are still more effective.

This book has another appealing feature—it is not snobbish or pompous! (Missing from its pages are the types of excesses commonly found in the evolutionary literature, where methods are often claimed to harness primordial forces of nature!) And yet the book is frankly ambitious. To make the book accessible to readers who may lack a background in optimization, Laguna and Martí start from ground zero. But they do not become trapped there, and the progress from rudimentary to more advanced material is handled in a way that enables the reader to produce state-of-the-art implementations without having to negotiate a maze of impenetrable explanations. Of particular value is the concluding chapter (Chapter 10) that summarizes what is known, what is not known, and what is suspected about advanced issues of methodology.

It is true that Scatter Search has for many years been a small and somewhat isolated island in the sea of evolutionary approaches. But due to its recent successes, this situation has now conspicuously changed. The Scatter Search book provides a series of enlightening application examples exposing the rapidly growing character of the field—and giving due credit to the inventive contributions of researchers who have been responsible for these applications. Ultimately, the promise of future advances of the method lies in the hands of those who produce such contributions, including those who are only now becoming introduced to Scatter Search for the first time. Laguna and Martí provide a useful framework not only for these beginners, but also for the "old pros" who may wish to explore further.

Fred Glover

Preface

We came up with the idea of writing this book in the spring of 2001 and wrote most of it in the summer of 2002 when Rafael Martí visited Boulder, Colorado thanks to a grant from the Ministerio de Educación, Cultura y Deporte of the Spanish Government.

Our work is intended to be a reference book for researchers and practitioners interested in gaining in-depth understanding of scatter search with the goal of expanding this methodology or applying it to solving difficult practical problems. The book could also be used to complement other materials in a graduate seminar on metaheuristic optimization. We believe that scatter search has great potential due to its flexibility and proven effectiveness. The scatter search framework can be adapted to tackle optimization problems in a variety of settings and we have tried to demonstrate this throughout the book. For example, we show adaptations that deal with several solutions representations, including continuous and binary variables as well as permutation vectors. We also address unconstrained and constrained problems to illustrate the malleability of the method.

To make the book accessible to a large audience, we begin with three tutorial chapters. Although there is some intentional repetition across tutorials, we recommend reading all three chapters before moving to the advanced topics. The tutorials introduce elements of scatter search in a succinct and illustrative way. This introduction is essential for the novice and might be helpfully reinforcing for the seasoned metaheuristic researcher and practitioner.

The advanced strategies discussed in Chapter 5 build upon the fundamental concepts introduced in the tutorial chapters. This chapter may be viewed as a springboard for launching research projects that aim at advancing the scatter search methodology. Some ideas in this chapter have been fully developed and implemented, but others deserve further exploration.

The book is self-contained and in Chapter 6, we devote a few pages to summarize tabu search concepts that are critically linked with scatter search. These methodologies share a common past and were conceived to complement each other rather than compete for supremacy in the world of metaheuristics. Their common past makes these methods depart from those that prefer to use popular analogies with nature. The nomenclature (and corresponding notation) in scatter search is direct and attempts to be descriptive. We are not concerned with finding a creative acronym or a "cute" connection with a living organism in order to make the methodology appealing or marketable. Through careful crafting of implementations and robust experimentation, we have established the effectiveness of scatter search and at the same time avoided the temptation of interpreting its behavior as one resembling a creature or natural phenomena.

The book includes a compact disc with the C code of all the implementations discussed throughout the chapters. We have tried to do our best in creating effective (and hopefully bug-free) practical implementations. However, we are confident that the interested reader is likely to find improved ways of applying scatter search to the problems that we use for illustration as well as others found in practice. We would be grateful to hear from those interested readers not only to improve future versions of this book but also to learn and advance our own understanding of a promising solution methodology.

Manuel Laguna

Rafael Martí

Acknowledgements

We are in debt to many people but in particular to two very good friends and colleagues: Fred Glover and Vicente Campos. Thank you for your contributions to this book and, most of all, thank you for your friendship.

Chapter 1

INTRODUCTION

But we're absolute beginners, with eyes completely open ...

David Bowie, Absolute Beginners (1986)

Scatter search (SS) is an evolutionary method that has been successfully applied to hard optimization problems. The fundamental concepts and principles of the method were first proposed in the 1970s and were based on formulations, dating back to the 1960s, for combining decision rules and problem constraints. The method uses strategies for search diversification and intensification that have proved effective in a variety of optimization problems. This book provides a comprehensive examination of the scatter search methodology from both conceptual and practical points of view. In the conceptual side, we discuss the origins and evolution of scatter search, arriving at a template that is the basis for most current implementations. In the practical side, we introduce C code that implements the search mechanisms that are discussed throughout the book. The C code can be used "as is" to replicate our practical illustrations or can be modified and adapted to other optimization problems of interest.

As discussed in Glover (1998 and 1999), the approach of combining existing solutions or rules to create new solutions originated in the 1960s. In the area of scheduling, researchers introduced the notion of combining rules to obtain improved local decisions. Numerically weighted combinations of existing rules, suitably restructured so that their evaluations embodied a common metric, generated new rules. The conjecture that information about the relative desirability of alternative choices is captured in different forms by different rules motivated this approach. The combination strategy was devised with the belief that this information could be exploited more effectively when integrated than when treated in isolation (i.e., when existing selection rules are selected one at a time). In general, the decision rules created from such combination strategies produced better empirical outcomes than standard applications of local decision rules. They also proved superior to a "probabilistic learning approach" that used stochastic selection of rules at different junctures, but without the integration effect provided by generating combined rules.

In integer and nonlinear programming, associated procedures for combining constraints were developed, which likewise employed a mechanism for creating weighted combinations. In this case, nonnegative weights were introduced to create new constraint inequalities, called surrogate constraints. The approach isolated subsets of constraints that were gauged to be most critical, relative to trial solutions based on the surrogate constraints. This critical subset was used to produce new weights that reflected the degree to which the component constraints were satisfied or violated.

The main function of surrogate constraints was to provide ways to evaluate choices that could be used to create and modify trial solutions. A variety of heuristic processes that employed surrogate constraints and their evaluations evolved from this foundation. As a natural extension, these processes led to the related strategy of combining solutions. Combining solutions, as manifested in scatter search, can be interpreted as the primal counterpart to the dual strategy of combining constraints.

Scatter search operates on a set of solutions, the *reference set*, by combining these solutions to create new ones. When the main mechanism for combining solutions is such that a new solution is created from the linear combination of two other solutions, the reference set may evolve as illustrated in Figure 1-1. This figure assumes that the original reference set of solutions consists of the circles labeled A, B and C. After a non-convex combination of reference solutions A and B, solution 1 is created. More precisely, a number of solutions in the line segment defined by A and B are created; however, only solution 1 is introduced in the reference set. (The criteria used to select solutions for membership in the reference set are discussed later.) In a similar way, convex and non-convex combinations of original and newly created reference solutions create points 2, 3 and 4. The complete reference set shown in Figure 1-1 consists of 7 solutions (or elements).

More precisely, Figure 1-1 shows a precursor form of the resulting reference set. Scatter search does not leave solutions in a raw form produced by its combination mechanism, but subjects candidates for entry into the reference set to heuristic improvement, as we elaborate in later chapters.

Unlike a "population" in genetic algorithms, the reference set of solutions in scatter search tends to be small. In genetic algorithms, two solutions are randomly chosen from the population and a "crossover" or combination mechanism is applied to generate one or more offspring. A typical population size in a genetic algorithm consists of 100 elements, which are randomly sampled to create combinations. In contrast, scatter search chooses two or more elements of the reference set in a systematic way with the purpose of creating new solutions. Since the combination process considers at least all pairs of solutions in the reference set, there is a practical need for keeping the cardinality of the set small. Typically, the reference set in scatter search has 20 solutions or less. In general, if the reference set consists of b solutions, the procedure examines approximately $(3b\text{-}7)b/2$ combinations of four different types, as described in Chapter 5. The basic type consists of combining two solutions; the next type combines three solutions, and so on and so forth.

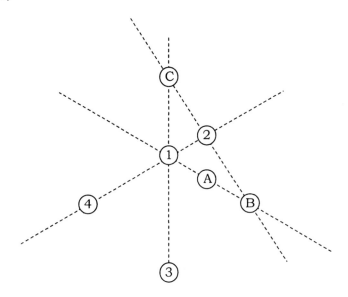

Figure 1-1. Basic reference set

Limiting the scope of the search to a selective group of combination types can be used as a mechanism for controlling the number of possible combinations in a given reference set. An effective means for doing this is to subdivide the reference set into "tiers" and to require that combined solutions must be based on including at least one (or a specified number) of the elements from selected tiers. The notion of dividing the reference set into multiple tiers and the development of rules to update such a set is explored in Section 1.2 of Chapter 5.

1.   HISTORICAL BACKGROUND

Scatter search (SS) was first introduced in 1977 (Glover 1977) as a heuristic for integer programming and it was based on strategies presented at a management science and engineering management conference held in Austin, Texas in September of 1967. The scatter search methodology has evolved since it was first proposed and hence we devote the rest of this section to the discussion of this evolution.

1.1   Original Proposal — 1977

The following is a slightly modified version of the scatter search description of scatter search in Glover (1977), which is the first published description of this methodology. In 1977, Glover described scatter search as a method that uses a succession of coordinated initializations to generate solutions. In this original proposal, solutions are purposely (i.e., non-randomly) generated to take account of characteristics in various parts of the solution space. Scatter search orients its explorations systematically relative to a set of reference points. Reference points may consist, for example, of the extreme points of a simplex obtained by truncating the linear programming basis cone, or of good solutions obtained by prior problem solving efforts.

The approach begins by identifying a convex combination, or weighted center of gravity, of the reference points. This central point, together with subsets of the initial reference points, is then used to define new sub-regions. Thereupon, analogous central points of the sub-regions are examined in a logical sequence (e.g., generally sweeping from smaller objective function values to larger ones in a minimization problem). Finally, these latter points are rounded to obtain the desired solutions. (Rounding, for any problems other than those with the simplest structures, should be either an iterative or a generalized adjacency procedure to accommodate interdependencies in the problem variables.)

The 1977 scatter search version can be visualized as depicted in Figure 1-2. Each of the points numbered 1 through 16 is the central point of an apparent sub-region of the simplex A, B and C. Point 8 is the center of (A, B, C) itself. In this example, A, B and C may not constitute the original reference points (which could, for example, have been 6, 7 and 11 or 4, 5, 12 and 13). The choice depends on the distribution of the points relative to each other and the feasible region in general. Thus, for example, it can be desirable to use envelopes containing derived reference points, and to use weights that bias central points away from centroids. When scatter search is carried out relative to reference points that lie on a single line, then it is reduced to a form of linear search.

In general, to keep the effort within desirable limits for larger dimensions, the points generated according to the example pattern may be examined in their indicated numerical sequence (or in an alternative sequence dictated by the slope of the objective function contour) until a convenient cutoff point is reached. Or one may examine a less refined collection of points.

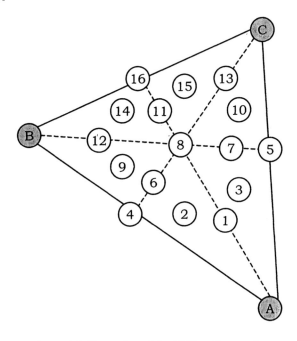

Figure 1-2. Illustration of the 1977 scatter search

The description terminates indicating that because scatter search may be applied to reference points obtained from a historical progression of solutions attempts and may also be used to influence this progression; the approach is conveniently suited to application with learning strategies. The following comments apply to this description of scatter search:

1. The approach focuses on combining more than two solutions to generate centroids. In fact, finding the centroid of all the reference points is considered a focal part of the search.

2. Generating solutions in a line, which is the most common strategy in modern scatter search implementation, is considered a reduced form of the more general method.

3. There is no reliance on randomization, but no guidelines are given to select "appropriate weights" to find biased centroids.

4. Combinations are described to be convex, although non-convex combinations are implicit in the illustrative example to make it possible to generate all the points in Figure 1-2 starting from points 6, 7 and 11.

5. The distribution of points relative to each other is considered important but no mechanism is discussed to encourage such dispersion.

To the best of our knowledge, scatter search was never applied or discussed again until 1990, when it was presented at the EPFL Seminar on Operations Research and Artificial Intelligence Search Methods (Lausanne, Switzerland). An article based on this presentation was published in 1994 (Glover 1994b).

1.2   Scatter / Tabu Search Hybrid — 1990

This description of scatter search provides implementation details that were not given in the original proposal. Also in Glover (1994b), the range of application is expanded to nonlinear (continuous) optimization problems, binary and permutation problems. The procedure is coupled with tabu search (TS), using forms of adaptive memory and aspiration criteria to influence the selection of points in a reference set consisting of several subsets. The method is described as one that generates "systematically dispersed set of points from a chosen set of reference points." The concept of weighted combinations is introduced as the main mechanism to generate new trial points on lines that join reference points. This version of scatter search emphasizes line searches and the use of weights to sample points from the line. The following specific elements are introduced:

1. The method starts from a set S of initial trial points from which all other sets derive. The generation of S is not proposed to be random.

2. The current reference set is defined as $R = H^* \cup S^*$, where H^* consists of h^* best solutions in H-T and S^* consists of the s^* best elements of S. H is the set of the best h solutions generated historically throughout the

search. T is a subset of solutions in H that are excluded from consideration.

3. An iteration consists of first generating the center of gravity for all the solutions in S^* and also the centers of gravity for all the subsets of size s^*-1 in S^*. These trial points are then paired with solution in R to create search lines for generating additional trial points. Elements of S^* are also paired with each other for the same purpose. Finally, the following solutions are also paired $H^* \times D(S^*)$ and $R \times D(S-S^*)$, where $D(X)$ denotes a diverse subset of set X.

4. When two solutions x and y are paired, trial solutions are generated with the line $z(w) = x + w(y - x)$, for $w = 1/2, 1/3, 2/3, -1/3$ and $4/3$.

5. When the quality of a trial solution z exceeds the average quality of the solutions in S^*, the search intensifies around z. The intensification consists of generating mid points of lines that join z with its nearest neighbors (i.e., trial points found during the line search that generated point z).

6. Adaptive memory of the recency and frequency type is used to control the admission of solutions to H and T.

In this version of scatter search, line searches start from centroids. That is, the centroids are generated as anchor points to initiate line searches using current and historical best solutions. The concept of combining high quality solutions with diverse solutions is also introduced. The method includes an intensification component, consisting of sampling more solutions from a line that has produced a "better-than-average" solution.

The concept of structured weighted combinations is also introduced to handle discrete optimization problems directly. This allows the representation of solutions that is natural to specific problems, such as permutations used in the context of traveling salesperson and job sequencing problems. Binary problems, like the knapsack, can also be handled with structured weighted combinations that are based on three properties:

1. *Representation property* — Each vector represents a set of votes for particular decisions (e.g., the decision of putting element j after element i in a permutation).

2. *Trial solution property* — A well-defined process translates set of votes prescribed by a vector into a trial solution to the problem of interest.

3. *Update property* — If a decision is made according to the votes of a given vector, a clearly defined rule exists to update all vectors for the residual problem so that both the representation property and the trial solution property continue to hold.

Several voting mechanisms with the aforementioned properties are suggested in connection with permutation problems as well as an adaptive procedure to dynamically change the associated weights. Although all these ideas were illustrated with several examples, including permutation and binary problems as well as a graph-partitioning problem, this version of scatter search has been implemented and tested only in the context of scheduling jobs on identical parallel machines in order to minimize total weighted tardiness (Mazzini 1998).

1.3   Scatter Search Template — 1998

In a way, the *scatter search template* (Glover 1998) is a simplification of the description in Section 1.2. This template, however, has served as the main reference for most of the scatter search implementations up to date. Specifically, the scatter search approach may be sketched as follows:

1. Generate a starting set of solution vectors to guarantee a critical level of diversity and apply heuristic processes designed for the problem considered as an attempt for improving these solutions. Designate a subset of the best vectors to be reference solutions. (Subsequent iterations of this step, transferring from Step 4 below, incorporate advanced starting solutions and best solutions from previous history as candidates for the reference solutions.) The notion of "best" in this step is not limited to a measure given exclusively by the evaluation of the objective function. In particular, a solution may be added to the reference set if the diversity of the set improves even when the objective value of such solution is inferior to other solutions competing for admission in the reference set.

2. Create new solutions consisting of structured combinations of subsets of the current reference solutions. The structured combinations are:

   a) chosen to produce points both inside and outside the convex regions spanned by the reference solutions.
   b) modified to yield acceptable solutions. (For example, if a solution is obtained by a linear combination of two or more solutions, a

generalized rounding process that yields integer values for integer-constrained vector components may be applied. Note that an acceptable solution may or may not be feasible with respect to other constraints in the problem.)

3. Apply the heuristic processes used in Step 1 to improve the solutions created in Step 2. (Note that these heuristic processes must be able to operate on infeasible solutions and may or may not yield feasible solutions.)

4. Extract a collection of the "best" improved solutions from Step 3 and add them to the reference set. The notion of "best" is once again broad; making the objective value one among several criteria for evaluating the merit of newly created points. Repeat Steps 2, 3 and 4 until the reference set does not change. Diversify the reference set, by re-starting from Step 1. Stop when reaching a specified iteration limit.

The first notable feature in this scatter search template is that its structured combinations are designed with the goal of creating weighted centers of selected subregions. This adds non-convex combinations that project new centers into regions that are external to the original reference solutions (see, e.g., solution 3 in Figure 1-1). The dispersion patterns created by such centers and their external projections have been found useful in several application areas.

Another important feature relates to the strategies for selecting particular subsets of solutions to combine in Step 2. These strategies are typically designed to make use of a type of clustering to allow new solutions to be constructed "within clusters" and "across clusters". Finally, the method is organized to use ancillary improving mechanisms that are able to operate on infeasible solutions, removing the restriction that solutions must be feasible in order to be included in the reference set.

The following principles summarize the foundations of the scatter search methodology as evolved to the form described in the template:

–   Useful information about the form (or location) of optimal solutions is typically contained in a suitably diverse collection of elite solutions.

–   When solutions are combined as a strategy for exploiting such information, it is important to provide mechanisms capable of constructing combinations that extrapolate beyond the regions spanned by the solutions considered. Similarly, it is also important to incorporate heuristic processes to map combined solutions into new solutions. The purpose of these combination mechanisms is to incorporate both diversity and quality.

– Taking account of multiple solutions simultaneously, as a foundation for creating combinations, enhances the opportunity to exploit information contained in the union of elite solutions.

The fact that the mechanisms within scatter search are not restricted to a single uniform design allows the exploration of strategic possibilities that may prove effective in a particular implementation. These observations and principles lead to the following template for implementing scatter search that consists of five methods.

1. A *Diversification Generation Method* to generate a collection of diverse trial solutions, using an arbitrary trial solution (or seed solution) as an input.

2. An *Improvement Method* to transform a trial solution into one or more enhanced trial solutions. (Neither the input nor the output solutions are required to be feasible, though the output solutions will more usually be expected to be so. If no improvement of the input trial solution results, the "enhanced" solution is considered to be the same as the input solution.)

3. A *Reference Set Update Method* to build and maintain a *reference set* consisting of the b "best" solutions found (where the value of b is typically small, e.g., no more than 20), organized to provide efficient accessing by other parts of the method. Solutions gain membership to the reference set according to their quality or their diversity.

4. A *Subset Generation Method* to operate on the reference set, to produce a subset of its solutions as a basis for creating combined solutions.

5. A *Solution Combination Method* to transform a given subset of solutions produced by the Subset Generation Method into one or more combined solution vectors.

The Diversification Generation Method suggested in this template emphasizes systematic generation of diversification over randomization. This theme remains unchanged from the original 1977 scatter search proposal. In the current version, however, an explicit reference to local search procedures is made. The Improvement Method expands the role of the mapping procedure originally suggested to transform feasible or infeasible trial solutions into "enhanced" solutions. Another significant change occurs in the updating of the reference set, which is greatly simplified when compared to the SS/TS version in Glover (1994b). The reference set in the template simply consists of the *b* best solutions found during the search. The Subset Generation Method is a departure from the original proposal. The method emphasizes the linear combination of two solutions. Combining more than two solutions is also suggested, but the lines searches are not "anchored" at the centroids of sub-regions (as suggested in the previous two versions).

The publication of this template for scatter search triggered the interest of researchers and practitioners, who translated these ideas into computer implementations for a variety of problems. A sample list of these applications appears in Table 1-1.

Table 1-1. Scatter search applications

Vehicle Routing	Rochat and Taillard (1995); Ochi, et al. (1998); Atan and Secomandi (1999); Rego and Leão (2000); Corberán, et al. (2002)
Traveling Salesperson Problem	Oliva San Martin (2000)
Arc Routing	Greistorfer (1999)
Quadratic Assignment	Cung et al. (1996)
Financial Product Design	Consiglio and Zenios (1999)
Neural Network Training	Kelly, Rangaswamy and Xu (1996); Laguna and Martí (2001)
Combat Forces Assessment Model	Bulut (2001)
Graph Drawing	Laguna and Martí (1999)
Linear Ordering	Laguna, Martí and Campos (2001)
Unconstrained Optimization	Fleurent, et al. (1996); Laguna and Martí (2000)
Bit Representation	Rana and Whitley (1997)
Multi-objective Assignment	Laguna, Lourenço and Martí (2000)
Optimizing Simulation	Glover, Kelly and Laguna (1996); Grant (1998)
Tree Problems	Canuto, Resende and Ribeiro (2001); Xu, Chiu and Glover (2000)

A short description (or vignette) for several representative applications of scatter search is provided in Chapter 8.

2. BASIC DESIGN

The scatter search methodology is very flexible, since each of its elements can be implemented in a variety of ways and degrees of sophistication. We have organized the presentation of the algorithmic details in such a way that the advanced features are progressively added to a basic design. It should be noted that the basic design is a search procedure quite capable of performing well in some settings. In this sense, implementing scatter search is not different than implementing other metaheuristic procedures. In tabu search, for example, it is customary to implement a basic procedure consisting of a short-term memory function and a simple aspiration level mechanism. If the basic implementation yields acceptable results, the advanced mechanisms of tabu search, such as long-term memory functions or strategic oscillation, are not included. The basic scheme, which includes all five methods outlined in Section 1-3 of this chapter, is depicted in Figure 1-3. This figure shows how the methods interact and puts in evidence the central role of the reference set.

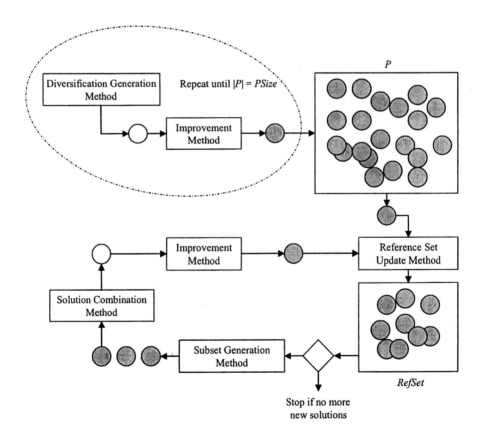

Figure 1-3. Schematic representation of a basic SS design

In Figure 1-3, the Diversification Generation and Improvement Methods are initially applied until the cardinality of P reaches *PSize* solutions that are different from each other. The darker circles represent improved solutions resulting from the application of the Improvement Method. The main search loop appears to the left of the box containing the reference solutions and labeled *RefSet*. The Subset Generation Method takes reference solutions as input to produce solution subsets to be combined. Solution subsets contain two or more solutions. The new trial solutions resulting from the application of the Combination Method are subjected to the Improvement Method and handed to the Reference Set Update Method. This method applies rules regarding the admission to the reference set of solutions coming from P or from the application of the Combination and Improvement Methods.

Of the five methods in the scatter search methodology, only four are strictly required. The Improvement Method is usually needed if high-quality outcomes are desired, but a scatter search procedure can be implemented without it. The advanced features of scatter search are related to the way the five methods are implemented. That is, the sophistication comes from the implementation of the SS methods instead of the decision to include or exclude some elements (like in the case of tabu search, as mentioned above). The basic scatter search procedure is shown in Figure 1-4.

---

1. Start with $P = \emptyset$. Use the <u>diversification generation</u> method to construct a solution and apply the <u>improvement</u> method. Let x be the resulting solution. If $x \notin P$ then add x to P (i.e., $P = P \cup x$), otherwise, discard x. Repeat this step until $|P| = PSize$.

2. Use the <u>reference set update</u> method to build $RefSet = \{ x^1, ..., x^b \}$ with the "best" b solutions in P. Order the solutions in $RefSet$ according to their objective function value such that x^1 is the best solution and x^b the worst. Make *NewSolutions* = TRUE.

while (*NewSolutions*) **do**

    3. Generate *NewSubsets* with the <u>subset generation</u> method. Make *NewSolutions* = FALSE.

    while (*NewSubsets* $\neq \emptyset$) **do**

        4. Select the next subset s in *NewSubsets*.

        5. Apply the <u>solution combination</u> method to s to obtain one or more new trial solutions x. Apply the <u>improvement</u> method to the trial solutions.

        6. Apply the <u>reference set update</u> method.

    if (*RefSet* has changed) **then**

        7. Make *NewSolutions* = TRUE.

    end if

        8. Delete s from *NewSubsets*.

    end while

end while

Figure 1-4. Basic scatter search pseudo-code

The basic procedure in Figure 1-4 starts with the creation of an initial reference set of solutions (*RefSet*). The Diversification Generation Method is used to build a large set P of diverse solutions. The size of P (*PSize*) is typically at least 10 times the size of *RefSet*. The initial reference set is built according to the Reference Set Update Method. For example, the Reference Set Update Method could consist of selecting b distinct and maximally diverse solutions from P. Alternatively, the method may be such that the best solutions from P, as measured by the objective function value, are selected as the reference set. (Chapter 5 explores several alternatives for implementing the Reference Set Update Method.)

Regardless of the rules used to select the reference solutions, the solutions in *RefSet* are ordered according to quality, where the best solution is the first one in the list. The search is then initiated by assigning the value of TRUE to the Boolean variable *NewSolutions*. In step 3, *NewSubsets* is constructed and *NewSolutions* is switched to FALSE. The simplest form of the Subset Generation Method consists of generating all pairs of reference solutions. That is, the method would focus on subsets of size 2 resulting in $(b^2-b)/2$ *NewSubsets*. The pairs in *NewSubsets* are selected one at a time in lexicographical order and the Solution Combination Method is applied to generate one or more trial solutions in step 5. These trial solutions are subjected to the Improvement Method, if one is available. The Reference Set Update Method is applied once again in step 6. The simplest form of the application of the Reference Update Method in this step is to build the new *RefSet* with the best solutions, according to the objective function value, from the current *RefSet* and the set of trail solutions. If *RefSet* changes after the application of the reference set update method the *NewSolutions* flag is switched to TRUE in step 7, indicating that at least one new solution has been inserted in the reference set. The subset s that was just subjected to the Combination Method is deleted from *NewSubsets* in step 8.

The basic procedure in Figure 1-4 terminates after all subsets in *NewSubsets* are subjected to the combination method and none of the improved trial solutions are admitted to *RefSet* under the rules of the Reference Set Update Method. Hence, the strategies for updating the reference set embedded in the Reference Set Update Method are responsible for the total searching time.

2.1   Summary of Notation

The notation introduced in connection with the basic scatter search in Figure 1-4 will be the basis for further developments. This is why we pause to summarize the notation before introducing more material.

RefSet	Reference set of solutions
b	Size of the reference set (i.e., $\|RefSet\| = b$)
x^i	The ith solution in the reference set. Since the reference set is typically ordered according to solution quality, x^1 is the best solution in *RefSet* and x^b is the worst.
P	Set of solutions generated with the Diversification Generation Method. This set is also referred to as "population" but it does not have the same meaning as in genetic algorithms.
PSize	Size of the population *P* of diverse solutions (i.e., $\|P\| = PSize$)
NewSubsets	List of subsets of reference solutions that are subjected to the Combination Method. The size of the list and the number of reference solutions in each subset is determined by the Subset Generation Method.
s	A subset of reference solutions. For instance, $s = \{x^1, x^2\}$ is the subset that contains the two best solutions in *RefSet*.
s_{size}	The size of the subset of reference solutions.
NewSolutions	A Boolean variable that indicates whether or not a new solution has become a member of *RefSet*. Since the Subset Generation Method typically generates only those subsets that include at least one new solution, the *NewSolutions* variable is used to decide whether to apply the Subset Generation Method to a *RefSet*.

Some advanced designs use the following additional notation:

$RefSet_i$	This is used when the reference set is partitioned into tiers. For example, a typical partition consists of two tiers: the high-quality tier and the diverse tier. The tiers are such that $RefSet = \bigcup_i RefSet_i$ and $\bigcap_i RefSet_i = \varnothing$. We use this notation in Chapter 5.

b_i Size of the ith tier in the reference set (i.e., $|RefSet_i| = b_i$ and $b = \sum_i b_i$)

$d(x,y)$ Dissimilarity measure between solution x and solution y. For example, Euclidean distances are typically used to measure the dissimilarity of two solutions that are represented by a vector of real numbers.

$d_{min}(x)$ Minimum dissimilarity between solution x and a set of other solutions under consideration. This is used, for instance, to represent the minimum dissimilarity from a solution $x \in P$ and all solutions $y \in RefSet$.

Additional notation will be introduced, as needed it, in later chapters, where details of SS implementations are discussed.

3.   C CODE CONVENTIONS

This book includes computer code written in C, which implements the methods and search strategies introduced in each chapter. Although we assume that the reader is familiar with the C language, the programs are such that anyone with working knowledge of computer programming would be able to follow the logic of the various computer codes shown throughout the book. We have adopted ANSI C as our standard with the goal of creating programs that are as portable and accessible as possible. We have tested our codes across different platforms, including those with the Windows system as well as Unix and Linux. Since some of the procedures require the generation of random numbers, our codes include a hardware-independent random number generator that makes the experiments reproducible in the tested platforms.

Although the complete codes are in the compact disc included with the book, we show code within the text to discuss implementation issues. These issues are relevant from the point of view of adapting the codes to situations that are different than the examples discussed in this book. Typographically, when programs are included in the text, they look like this:

```
#include "ss.h"

void SSCombine(n, x, s)
```

```
int m, r;
float *eval;
```

Our convention of handling all errors and exceptional cases is to call the SSAbort function. We call this function with text that briefly explains the type of error found. SSAbort, shown in Function 1-1, is located in file SSTools.c of the Chapter1 folder in the accompanying disc.

Function 1-1. SSAbort — File: SSTools.c

```
void SSAbort(char text[])
/* Prints an error message and terminates the program */
{
    printf("\n%s",text);
    exit(1);
}
```

Arrays in C are natively "zero-offset" and the index 0 is always considered. If a code contains the declaration "int array[3];" then the valid references are array[0], array[1] and array[2]. However, most of our algorithms use arrays indexed from 1 to *n*, so we have adapted the C language to the natural indexation of the methods instead of re-indexing the algorithms from 0 to *n*-1. Therefore, we always use unit-offset arrays in this book and if we say that an array has size or dimension equal to 3, the valid references will be array[1], array[2] and array[3]. For instance, the memory allocation of an array of integers is performed with SSInt_array, as shown in Function 1-2.

Function 1-2. SSInt_array — File: SSMemory.c

```
int *SSInt_array(int size)
/* Allocates an int array indexed from 1 to size */
{
    int *aux;

    aux = (int*)calloc(size,sizeof(int));
    if(!aux) SSAbort("Memory allocation problem: SSint_array");
    return aux-1;
}
```

Since a matrix can be viewed as an array of vectors, SSint_matrix performs the memory allocation of a matrix, by successively calling the SSint_array function.

Function 1-3. SSInt_matrix — File: SSMemory.c

```
int **SSInt_matrix(int nrows, int ncolumns)
/* Allocates an int matrix with "nrows" rows and "ncolumns"
columns */
{
    int i,**aux;

    aux = (int**)calloc(nrows,sizeof(int*));
    if(!aux) SSAbort("Memory allocation problem:SSInt_matrix");

    aux--;
    for(i=1;i<=nrows;i++)
    aux[i] = SSint_array(ncolumns);

    return aux;
}
```

Function 1-4 frees a matrix of integers previously allocated with SSint_matrix.

Function 1-4. SSFree_int_matrix — File: SSMemory.c

```
void SSFree_int_matrix(int **matrix, int nrows)
/* Frees an int matrix allocated with SSInt_matrix */
{
    int i;

    for(i=1;i<=nrows;i++)
      free(matrix[i]+1);
    free(matrix+1);
}
```

Memory allocation for arrays of other types, such as "double", is performed with functions similar to Functions 1-2 to 1-4. Specifically the following three functions are obtained by replacing the "int" type with the "double" type in Functions 1-2 to 1-4:

```
SSDouble_array
SSDouble_matrix
SSFree_double_matrix
```

The SSMemory.c file contains these memory allocation functions for the double data type. The SS.h file contains the prototypes of the functions as well as the definitions of several key data structures. The reference set and its associated values are stored in the REFSET structure (see Structure 1.1). This structure contains the size of the reference set (i.e., the value of b) and the value of the Boolean variable *NewSolutions*. For each solution i ($i = 1,..,b$) REFSET also contains:

- sol[i][j], the value of each variable j
- ObjVal[i], the objective function value
- iter[i], the iteration number in which solution i entered the reference set

The solution vector sol[i] may be viewed as the "i^{th} row" of the sol array. Solutions in the reference set are ordered according to their objective value and the location array order indicates the ranking of the solutions; that is, sol[order[1]] is the best solution, sol[order[2]] is the second best solution and sol[order[b]] is the worst solution in the reference set.

Structure 1.1. REFSET — File: SS.h

```
typedef struct REFSET
{
    int b;             /* Size                            */
    double **sol;      /* Solutions                       */
    double *ObjVal;    /* Objective value of solutions    */
    int *order;        /* Order of solutions              */
    int *iter;         /* Entering iteration              */
    int NewSolutions;  /* =1 if new element has been added */
} REFSET;
```

In addition, the SS.h file contains the definition of the P structure (Structure 1.2), which uses the sol matrix to store the solutions in the population. The objective function values are stored in the ObjVal array and the number of solutions is stored in the PSize variable.

Structure 1.2. P — File: SS.h

```
typedef struct P
{
    int PSize;        /* Size of the P set             */
    double **sol;     /* Solutions in the set          */
    double *ObjVal;   /* Objective value of solutions  */
} P;
```

The SS structure (Structure 1.3) contains pointers to the REFSET and P structures. The number of variables in the problem is stored in nvar. In problems with continuous variables, each variable *j* is considered to be bounded and the bounds are stored in low[j] and high[j].

Structure 1.3. SS — File: SS.h

```
typedef struct SS
{
    int nvar;           /* Number of variables in the problem */
    double *high;       /* Upper bounds                       */
    double *low;        /* Lower bounds                       */

    REFSET *rs;
    P      *p;
} SS;
```

All our scatter search implementations begin with a call to the SSProblem_Definition function, which initializes the search process (see Function 1-5). This function allocates memory for the SS structure and sets up the search with the specified parameter values. The function has the following arguments: nvar, b and PSize. The function returns a pointer to the allocated SS structure.

Function 1-5. SSProblem_Definition — File: SSMemory.c

```
SS *SSProblem_Definition(int nvar,int b,int PSize)
{
    SS *pb;

    pb = (SS*) calloc(1,sizeof(SS));
    if(!pb) SSAbort("Memory allocation problem: SS");

    pb->nvar = nvar;

    /* Diverse Set P */
    pb->p = (P*) calloc(1,sizeof(P));
    if(!pb->p) SSAbort("Memory allocation problem: P");
    pb->p->Psize    = PSize;
    pb->p->sol      = SSDouble_matrix(PSize,nvar);
    pb->p->ObjVal   = SSDouble_array(PSize);

    /* Reference Set */
    pb->rs = (REFSET*) calloc(1,sizeof(REFSET));
```

```
if(!pb->rs) SSabort("Memory allocation problem: REFSET");
pb->rs->b      = b;
pb->rs->sol    = SSDouble_matrix(b,nvar);
pb->rs->ObjVal = SSDouble_array(b);
pb->rs->order  = SSInt_array(b);
pb->rs->iter   = SSInt_array(b);

return pb;
}
```

The counterpart of the problem definition function SSProblem_Definition is the SSFree_DataStructures function that frees the allocated memory. Function 1-6 shows a sample main function that uses the general structure that starts with a call to SSProblem_Definition and ends with a call to SSFree_DataStructures.

Function 1-6. Main — File: SSMain.c

```
int main(void)
{
    SS *pb;              /* Pointer to problem data    */
    int nvar  =  10;     /* Number of variables        */
    int b     =  10;     /* Size of reference set      */
    int PSize = 100;     /* Size of P                  */

    pb = SSProblem_Definition(nvar,b,PSize);

    /* Insert here the Scatter Search code */

    SSFree_DataStructures(pb);
    return 0;
}
```

The computer codes in this section can be found in the Chapter1 folder of the compact disc accompanying this book.

Chapter 2

TUTORIAL
Unconstrained Nonlinear Optimization

The answer is out there ... It's looking for you. And, it will find you, if you want it to.

Carrie-Anne Moss in The Matrix (2000)

In the following three chapters, we describe the development of both basic and somewhat advanced scatter search procedures for unconstrained nonlinear optimization, 0-1 knapsack problems and linear ordering problems. These illustrative examples give us the opportunity to discuss practical implementation issues associated with problems from different settings. The procedures in these chapters were conceived with the purpose of illustrating the implementation of the scatter search methodology and not with the purpose of competing with the best optimization technology available for the solution of the chosen problems. For example, it is well-known that exact procedures exist to optimally solve fairly large instances of the knapsack problem. However, in the case of the linear ordering problem, the procedure described in Chapter 4 is based on a state-of-the-art scatter search implementation reported in the literature.

As discussed in Chapter 1, scatter search consists of 5 methods. Some of these methods, such as the one used to update the reference set, can be employed in scatter search implementations designed in the context of entirely different problems. In particular, the same Reference Set Update Method can be used in a scatter search implementation for problems with either binary variables or continuous variables. However, the Combination Method is usually context-specific and therefore a method to combine solutions represented by continuous or binary variables is not necessarily the most effective in a setting where solutions are represented by permutations.

We start our series of tutorials with the description of a scatter search implementation for the solution of unconstrained nonlinear optimization problems in a continuous space. The problem is mathematically represented as follows:

Minimize $f(x)$
Subject to $l \leq x \leq u$

where $f(x)$ is a non-linear function of x, and x is a vector of continuous and bounded variables. For the purpose of providing detailed descriptions in this tutorial, we apply our design to the following problem instance:

Minimize
$$100\left(x_2 - x_1^2\right)^2 + \left(1 - x_1\right)^2 + 90\left(x_4 - x_3^2\right)^2 + \left(1 - x_3\right)^2 +$$
$$10.1\left(\left(x_2 - 1\right)^2 + \left(x_4 - 1\right)^2\right) + 19.8\left(x_2 - 1\right)\left(x_4 - 1\right)$$

Subject to $-10 \leq x_i \leq 10$ for $i = 1, \ldots, 4$

This problem is "Test Case #6" in (Michalewicz and Logan, 1994). The best solution found by Michalewicz and Logan has an objective function value of 0.001333, which was reportedly found after 500,000 runs of a genetic algorithm (GA)[1].

1.   DIVERSIFICATION GENERATION METHOD

Our illustrative Diversification Generation Method employs controlled randomization and frequency-based memory to generate a set of diverse solutions. The use of frequency-based memory is typical in implementations of tabu search. Frequency-based memory is particularly useful when implementing long-term memory strategies, as described in Chapter 4 of Glover and Laguna (1997). (We discuss additional uses of memory in scatter search in Chapter 6.) In our current implementation, we start by dividing the range of each variable $u_i - l_i$ into 4 sub-ranges of equal size. Then, a solution is constructed in two steps. First a sub-range is randomly selected. The probability of selecting a sub-range is inversely proportional to its frequency count. Then a value is randomly generated within the selected sub-range. The number of times sub-range j has been chosen to generate a value for variable i is accumulated in *freq*(i, j).

---

[1]   A single run of the GA in Michalewicz and Logan (1994) consists of 1000 iterations. An iteration requires 70 evaluations of the objective function, where 70 is the size of the population. Therefore, this particular GA implementation required 35 billion objective function evaluations to find this solution.

The Diversification Generation Method is used at the beginning of the search to generate a set P of *PSize* diverse solutions. In most scatter search applications, the size of P is generally set at max(100, 5*b), where b is the size of the reference set. Although the actual scatter search implementation generates 100 solutions for this example, Table 2-1 only shows the first ten.

The 100 solutions generated by the computer code in Section 2.1 are diverse with respect to the values that each variable takes in each of the sub-ranges. Note, however, that the generation is done without considering the objective function. In other words, the Diversification Generation Method focuses on diversification and not on the quality of the resulting solutions, as evident from the objective function values reported in Table 2-1. In fact, the objective function values for the entire set of 100 solutions range from 1,689.7 to 1,542,078 with an average of 392,033. These objective function values are very large considering that the optimal solution to this problem has an objective function value of zero.

Table 2-1. Diverse solutions

Solution	x_1	x_2	x_3	x_4	$f(x)$
1	1.11	0.85	9.48	-6.35	835546.2
2	-9.58	-6.57	-8.81	-2.27	1542078.9
3	7.42	-1.71	9.28	5.92	901878.0
4	8.83	-8.45	4.52	3.18	775470.7
5	-6.23	7.48	6	7.8	171450.5
6	1.64	3.34	-8.32	-8.66	546349.8
7	0.2	-3.64	-5.3	-7.03	114023.8
8	-3.09	6.62	-2.33	-3.12	7469.1
9	-6.08	0.67	-6.48	1.48	279099.9
10	-1.97	8.13	-5.63	8.02	54537.2

To verify that the Diversification Generation Method is operating as expected, we show in Table 2-2 the frequency values corresponding to the complete set of 100 solutions.

Table 2-2. Frequency counts

Range	x_1	x_2	x_3	x_4
-10 to 5	19	25	26	29
-5 to 0	25	18	22	21
0 to 5	26	27	29	25
5 to 10	30	30	23	25

Note that the frequency counts in Table 2-2 are very similar for each range and variable, with a minimum value of 18 and a maximum value of

30. The expected frequency value for each range is 25 for a set of 100 solutions. This value is observed four times in Table 2-2.

1.1   Computer Code

The SSGenerate_Sol function in SSP1.c file of the Chapter2 folder implements the Diversification Generation Method, and returns a solution each time that the function is called. The frequency count for each variable and sub-range is stored in the freq matrix allocated in the SS structure pb, which is defined in the SS1.h file. The frequency count data is used to first select a sub-range and then choose a value within the sub-range. This process is repeated for each variable until all the variables have been assigned a value. The sum variable accumulates the total frequency and the rfreq array stores the "frequency complement", that is, the difference between the total frequency and the frequency of each sub-range. These complementary values are accumulated in the total variable. The total value and the values stored in rfreq are used to probabilistically select the sub-range and store it in the index variable. Finally, the index variable is used to update the sub-range frequency and to draw a sample from the selected sub-range.

Function 2-1. SSGenerate_Sol — File: SSP1.c

```
void SSGenerate_Sol(SS *pb, double sol[])
{
    int i,j,var,total,sum,index,*rfreq;
    double r,value,range;

    rfreq = SSInt_array(4);
    for(var=1;var<=pb->nvar;var++)
    {
        /* Compute frequency complements */
        sum=total=0;
        for(i=1;i<=4;i++)    sum += pb->freq[var][i];
        for(i=1;i<=4;i++)    {
            rfreq[i] = sum - pb->freq[var][i];
            total    += rfreq[i];
        }
        /* Choose a sub-range (index) */
        if(total==0)    index = SSGetrandom(1,4);
        else {
            j = SSGetrandom(1,total);
            index=1;
```

```
        while(j>rfreq[index])
            j -= rfreq[index++];
    }
    if(index>4) SSAbort("Failure in SSGenerate_Sol");
    /* Update freq */
    pb->freq[var][index]++;

    /* Select a value in the index sub-range */
    range= pb->high[var]-pb->low[var];
    r = SSRandNum();
    value=pb->low[var]+(index-1)*(range/4) + (r*range/4);
    sol[var]=value;
    }
    free(rfreq+1);
}
```

Function 2-2, SSCreate_P, in the SSP1.c file creates a population of PSize elements by successively calling function SSGenerate_Sol. Upon return from SSGenerate_Sol, the newly generated solution is improved with a call to SSimprove_solution. Improved solutions are examined to determine whether they are new or they already belong to the P set. If the improved solution is new, it is added to P; otherwise the solution is discarded. The function ends when PSize different improved solutions have been added to P.

Function 2-2. SSCreate_P — File: SSP1.c

```
void SSCreate_P(SS *pb)
{
    double *sol,obj_val;
    int currentPSize,j,equal;

    sol = SSDouble_array(pb->nvar);
    currentPSize=1;
    while(currentPSize<=pb->p->PSize)
    {
        SSGenerate_Sol(pb,sol);
        obj_val = sol_value(sol);
        SSImprove_solution(pb,sol,&obj_val);

        /* Check whether sol is a new solution */
        j=1;equal=0;
        while(j<currentPSize && !equal)
```

```
     equal=SSEqualSol(sol,pb->p->sol[j++],pb->nvar);

   /* Add improved solution to P */
   if(!equal) {
     for(j=1;j<=pb->nvar;j++)
         pb->p->sol[currentPSize][j] = sol[j];
     pb->p->ObjVal[currentPSize++] = obj_val;
   }
  }
  free(sol+1);
}
```

2.   IMPROVEMENT METHOD

Since we are developing a solution procedure for a class of unconstrained optimization problems and the Diversification Generation Method constructs solutions by drawing values within the allowed bounds for each decision variable, these solutions are guaranteed to be feasible. This means that in this case the Improvement Method always starts from a feasible solution. In general, however, Improvement Methods must be capable of handling starting solutions that are either feasible or infeasible. We will address this issue in Chapter 3, where we deal with a class of constrained optimization problems.

Table 2-3. Improved solutions

Solution	x_1	x_2	x_3	x_4	$f(x)$
1	2.5	5.4	2.59	5.67	1002.7
2	-0.52	-0.5	0.35	-0.14	138.5
3	-2.6	5.9	4.23	10	7653.7
4	0.49	0.53	2.47	5.89	213.7
5	-3.04	9.45	1.14	0.41	720.1
6	-1.4	2.46	0.37	-3.94	1646.7
7	-0.36	-0.31	0.8	1.14	57.1
8	-1.63	2.51	0.73	0.56	21.5
9	-0.8	0.69	-1.16	1.5	11.2
10	-2.47	5.32	-2.92	8.15	1416.7

The Improvement Method we use in this tutorial consists of a classical local optimizer for unconstrained non-linear optimization problems. In particular, we apply Nelder and Mead's simplex method (Nelder and Mead, 1965) to each solution in Table 2-1. This method requires an input parameter to specify the number of objective function evaluations. Since our goal for this tutorial chapter is to illustrate the scatter search methodology instead of finding solutions of the highest quality possible, we limit the number of objective function evaluations to 50. After the application of this local optimizer, the solutions in Table 2-1 are transformed to the solutions in Table 2-3.

The objective function values now range between 11.2 and 7,653.7 for the improved solutions in Table 2-3. It is interesting to point out that if we increase the number of objective function evaluations in the Improvement Method to 500, the objective function values of the 100 solutions in P would range between 0.00 and 7.66, with an average of 0.38. Also, 56 solutions out of the 100 would converge to the same local minimum, thus effectively reducing the cardinality of P. As shown in step 1 of the basic SS procedure of Figure 1-3, we could apply the Diversification Generation Method again until *PSize* different solutions are found after executing the Improvement Method. Incidentally, the convergence point for those 56 solutions turns out to be the global optimum that sets $x = (1, 1, 1, 1)$ for an objective function value of zero. Hence for this small example, a single application of the Diversification Generation and Improvement methods with a moderate limit on the number of objective function evaluations is sufficient to find the optimal solution. Nonetheless, in order to show the performance of all the scatter search elements, we keep the number of evaluations set to 50 in the Improvement Method.

2.1   Computer Code

The SSImprove_solution function has three arguments: the pb pointer to the SS structure where all relevant problem information is stored, the solution to be improved (sol), and the corresponding solution value. Both, the solution and its value are passed as pointers to the memory locations where the improved values are stored after the function returns.

Given a solution sol, the Nelder and Mead method starts by perturbing each value sol[i] to create an initial simplex from which to begin the local search. The simplex matrix stores the perturbed solutions and the values array stores their corresponding objective function values. The SS_Simplex function is called to perform the search starting from the initial simplex. Function 2-3 calls SS_Simplex with a limit of 50 objective function evaluations. SS_Simplex returns a new simplex on which one vertex contains the best solution found.

Function 2-3. SSImprove_solution — File: SSImprove1.c

```
void SSImprove_solution(SS *pb,double sol[],double *value)
{
```

```
double **simplex,*values;
int i,j,best_sol;

simplex = SSdouble_matrix(pb->nvar+1,pb->nvar);
values  = SSDouble_array(pb->nvar+1);

/* Copy original solution in nvar+1 vertices */
for(j=1;j<=pb->nvar+1;j++)
for(i=1;i<=pb->nvar;i++)
   simplex[j][i]=sol[i];
values[1]=*value;

/* Perturb nvar vertices */
for(j=1;j<=pb->nvar;j++)   {
   simplex[j+1][j] += 0.1*(pb->high[j]-pb->low[j]);
   if(simplex[j+1][j]>pb->high[j])
      simplex[j+1][j]=pb->high[j];
   if(simplex[j+1][j]<pb->low[j])
      simplex[j+1][j]=pb->low[j];
   values[j+1] = sol_value(simplex[j+1]);
}
/* Call the Simplex method for 50 evaluations */
SS_Simplex(pb,simplex,values,50);

/* Select the best vertex */
best_sol=0;
for(i=1;i<=pb->nvar+1;i++)
   if( *value>values[i]) {
      *value=values[i]; best_sol=i;
   }
if(best_sol>0) {
   for(i=1;i<=pb->nvar;i++)   {
      sol[i]=simplex[best_sol][i];
      if(sol[i]<pb->low[i])    sol[i]=pb->low[i];
      if(sol[i]>pb->high[i])   sol[i]=pb->high[i];
   }
   *value = sol_value(sol);
}
free(values+1); SSFree_double_matrix(simplex,pb->nvar+1);
}
```

Although we don't discuss the computer code for ss_simplex, it can be found in the SSImprove1.c file of the Chapter2 folder included in the accompanying compact disc.

3.   REFERENCE SET UPDATE METHOD

The reference set, *RefSet*, is a collection of both high quality solutions and diverse solutions that are used to generate new solutions by way of applying the Combination Method. In this example, we illustrate the implementation of a simple mechanism to construct an initial reference set and then update it during the search. The size of the reference set is denoted by $b = b_1 + b_2 = |RefSet|$. The construction of the initial reference set starts with the selection of the best b_1 solutions from P. These solutions are added to *RefSet* and deleted from P. (Recall that P is the set of solutions constructed with the application of the Diversification Generation Method and the Improvement Method. All solutions in P are different from each other.)

For each solution in P-*RefSet*, the minimum of the Euclidean distances to the solutions in *RefSet* is computed. Then, the solution with the maximum of these minimum distances is selected. This solution is added to *RefSet* and deleted from P and the minimum distances are updated. This process is repeated b_2 times, where $b_2 = b - b_1$. The resulting reference set has b_1 high-quality solutions and b_2 diverse solutions.

Let us apply this initialization procedure to our example, using the following parameter values: $b = 10$, $b_1 = 5$ and $b_2 = 5$. Table 2-4 shows the best b_1 solutions in P, which are immediately added to the *RefSet*. The first column in this table shows the solution number in P, followed by the variable values and the objective function value. Therefore, solution 35 in P has the best objective function value and becomes the first solution in *RefSet*.

Table 2-4. High quality solutions in *RefSet*

Solution number in P	x_1	x_2	x_3	x_4	$f(x)$
35	-0.0444	0.0424	1.3577	1.8047	2.1
46	1.133	1.2739	-0.6999	0.5087	3.5
34	-0.0075	0.0438	1.4783	2.2693	3.5
49	1.1803	1.4606	-0.344	0.2669	5.2
38	1.0323	0.9719	-0.8251	0.695	5.3

We then calculate the minimum distance $d_{min}(x)$ between each solution x in P-*RefSet* and the solutions y currently in *RefSet*. That is,

$$d_{min}(x) = \underset{y \in RefSet}{Min} \{d(x,y)\},$$

where $d(x,y)$ is the Euclidean distance between x and y. Mathematically:

$$d(x,y) = \sqrt{\sum_{i=1}^{n}(x_i - y_i)^2}$$

For example, the minimum distance between solution 3 in Table 2-3 (i.e., x^3) and the *RefSet* solutions in Table 2-4 (i.e., x^{35}, x^{46}, x^{34}, x^{49} and x^{38}) is calculated as follows:

$$d_{min}(x^3) = Min\{d(x^3,x^{35}), d(x^3,x^{46}), d(x^3,x^{34}), d(x^3,x^{49}), d(x^3,x^{38})\}$$
$$d_{min}(x^3) = Min\{10.780, 12.235, 10.407, 12.232, 12.231\} = 10.407$$

The maximum d_{min} value for the solutions in *P-RefSet* corresponds to solution 37 in P ($d_{min}(x^{37}) = 13.59$). We add this solution to *RefSet*, delete it from P and update the d_{min} values. The new maximum d_{min} value of 12.64 corresponds to solution 30 in P, so the diverse solutions added to *RefSet* are shown in Table 2-5.

Table 2-5. Diverse solutions in *RefSet*

Solution	x_1	x_2	x_3	x_4	$f(x)$
37	-3.4331	10	1.0756	0.3657	1104.1
30	3.8599	10	-4.0468	10	9332.4
45	-4.4942	10	3.0653	10	13706.1
83	-0.2414	-6.5307	-0.9449	-9.4168	17134.8
16	6.1626	10	0.1003	0.1103	78973.2

After the initial reference set is constructed, the Combination Method is applied to the subsets generated as outlined in Section 4 of this chapter. We use the so-called *static update* of the reference set after the application of the Combination Method. Trial solutions that are constructed as combination of reference solutions are placed in a solution pool, denoted by *Pool*. After the application of both the Combination Method and the Improvement Method, the *Pool* is full and the reference set is updated. The new reference set consists of the best b solutions from the solutions in the current reference set and the solutions in the pool, i.e., the update reference set contains the best b solutions in *RefSet* $\cup$ *Pool*. Table 2-6 shows this new reference set.

Table 2-6. RefSet after update

x_1	x_2	x_3	x_4	$f(x)$
1.1383	1.2965	0.8306	0.715	0.14
0.7016	0.5297	1.2078	1.4633	0.36
0.5269	0.287	1.2645	1.6077	0.59
1.1963	1.3968	0.6801	0.446	0.62
0.3326	0.1031	1.3632	1.8311	0.99
0.3368	0.1099	1.3818	1.9389	1.02
0.3127	0.0949	1.3512	1.8589	1.03
0.7592	0.523	1.3139	1.7195	1.18
0.2004	0.0344	1.4037	1.9438	1.24
1.3892	1.9305	0.1252	-0.0152	1.45

If the reference set remains unchanged after the updating procedure, a rebuilding step is performed. In this example, the rebuilding consists of keeping the best b_1 solutions in *RefSet* and using the Diversification Generation Method to put diverse solutions in the place of the worst b_2 solutions in the current reference set.

3.1   Computer Code

Function 2-4, SSCreate_RefSet, creates the reference set from the solutions currently in *P*. First, the elements in *P* are ordered according to their objective function values, where the position of the best element is given by p_order[1], the second best by p_order[2], and so on. Then, the best b_1 elements are copied from *P* to *RefSet*. The order of the elements in *RefSet* is stored in the order array of the REFSET data structure defined in the SS1.h include file. This data structure is part of the problem data structure and therefore is accessed using the pointer pb->rs. The iteration number in which solution i is added to the *RefSet* is stored in iter[i] pointed by pb->rs. Since the initial *RefSet* is created in the first iteration then iter[i] takes on the value of 1 for all solutions.

For each solution i in *P*, the distance to *RefSet* (minimum of the distances to the solutions in *RefSet*) is computed with the SSDist_RefSet function and stored in min_dist[i]. Then, SSMax_dist_index identifies the solution with the maximum distance measure. This solution becomes the one that is added to *RefSet*. Once it has been added, the SSUpdate_distances function updates the distances for the remaining elements in *P* to take into consideration the solution most recently added to the reference set. Finally, NewSolutions, the Boolean variable that indicates whether or not a new solution has become a member of *RefSet*, is set to 1 (i.e., True).

Function 2-4. SSCreate_RefSet — File: SSRefSet1.c

```
void SSCreate_RefSet(SS *pb)
{
    double *min_dist;
    int b1,i,j,a,*p_order,*rs_order;

    /* Order solutions in P */
    p_order = SSOrder(pb->p->ObjVal, pb->p->PSize, pb->Opt);

    /* Add the b1 best solutions in P to RefSet */
    b1 = pb->rs->b / 2;
    for(i=1;i<=b1;i++)
    {
        for(j=1;j<=pb->nvar;j++)
            pb->rs->sol[i][j] = pb->p->sol[p_order[i]][j];
        pb->rs->ObjVal[i] = pb->p->ObjVal[p_order[i]];
        pb->rs->iter[i]   = 1;
        pb->rs->order[i]  = i;
    }

    /* Compute minimum distances from P to RefSet */
    min_dist = SSDouble_array(pb->p->PSize);
    for(i=1;i<=pb->p->PSize;i++)
        min_dist[i]= SSDist_RefSet(pb,b1,pb->p->sol[i]);

    /* Add the diverse b-b1 solutions to RefSet */
    for(i=b1+1;i<=pb->rs->b;i++)
    {
        a=SSMax_dist_index(pb,min_dist);

        for(j=1;j<=pb->nvar;j++)
            pb->rs->sol[i][j] = pb->p->sol[a][j];
        pb->rs->ObjVal[i] = pb->p->ObjVal[a];
        pb->rs->iter[i]   = 1;

        SSUpdate_distances(pb,min_dist,i);
    }

    /* Compute the order in RefSet */
    rs_order = SSOrder(pb->rs->ObjVal,pb->rs->b,pb->Opt);
    for(i=1;i<=pb->rs->b;i++)
```

```
        pb->rs->order[i] = rs_order[i];

    pb->rs->NewSolutions = 1;
    pb->CurrentIter      = 1;

    free(rs_order+1);free(p_order+1);free(min_dist+1);
}
```

Function 2-4 calls several functions that we don't describe in detail here but that there are part of the codes distributed in the compact disc that accompanies this book. Specifically, the functions are:

- SSOrder, orders solutions by their objective function value
- SSDouble_array, allocates an array of doubles
- SSDist_RefSet, calculates the minimum distance between the given solution and the specified solutions in the reference set
- SSMax_dist_index, returns the index of the solution in *P* that has a maximum minimum distance to the solutions in *RefSet*
- SSUpdate_distances, updates the distance measures once a new solution has been added to *RefSet*

All of these functions are located in the SSTools.c file of the Chapter2 folder with the exception of the second function which can be found in SSMemory1.c along with all other memory allocation functions used in this tutorial.

Updating is another important operation related to the reference set. The update is performed with the SSUpdate_RefSet function, which appears below under the heading Function 2-5. This function first calls the SSCombine_RefSet function to execute the Combination Method, which generates new trial solutions and stores them in the pool matrix of size given by the value of pool_size. The pb->CurrentIter iteration counter is incremented by one unit to keep track of the iteration number in which the current solutions in the reference set were combined.

The Improvement Method discussed in Section 2 (i.e, SSImprove_solution) is applied to all the solutions in pool. Finally, SSTryAdd_RefSet attempts to add the solutions in pool to the current reference set. If the solution is better than the worst reference solution (i.e., pb->rs->sol[worst_index]) then the function adds the new trial solution and deletes the reference solution that is being replaced. The new trial solution is added to the reference set in the position that preserves the order that establishes that the first solution is the best and the last the worst.

Function 2-5. SSUpdate_RefSet — File: SSRefSet1.c

```
void SSUpdate_RefSet(SS *pb)
{
    int a;
    double value;

    pb->rs->NewSolutions=0;
    SSCombine_RefSet(pb);
    pb->CurrentIter++;

    for(a=1;a<=pb->pool_size;a++)
    {
      value=sol_value(pb->pool[a]);
      SSImprove_solution(pb,pb->pool[a],&value);
      SSTryAdd_RefSet(pb,pb->pool[a],value);
    }
    pb->pool_size=0;
}
```

Rebuilding is the third key operation associated with the reference set. Function 2-6, SSRebuild_RefSet, implements a mechanism to partially rebuild the *RefSet* when none of the new trial solutions generated with the Combination Method qualify to be added to the reference set. The method is similar to that one used to create the initial *RefSet* in the sense that it uses the min-max distance criterion for selecting diverse solutions. The rebuilding mechanism first constructs a new population P with a call to the SSCreate_P function. Then, the distance from each solution i in P to the first (best) b_1 solutions in *RefSet* is computed with SSDist_RefSet and stored in min_dist[i]. The solutions in P with the maximum minimum distance are added to the *RefSet*, thus replacing the worst b_2 solutions. The Boolean variable pb->rs->NewSolutions, which signals the presence of new solutions in the reference set, is then switched to 1 (i.e., True).

Function 2-6. SSRebuild_RefSet — File: SSRefSet1.c

```
void SSRebuild_RefSet(SS *pb)
{
    double *min_dist;
    int b1,i,j,a,index,*rs_order;

    b1 = pb->rs->b / 2;

    /* Create a new set P */
```

```
SSCreate_P(pb);

/* Compute minimum distances from P to RefSet */
min_dist = SSDouble_array(pb->p->PSize);
for(i=1;i<=pb->p->PSize;i++)
  min_dist[i]= SSDist_RefSet(pb,b1,pb->p->sol[i]);

/* Add the diverse b-b1 solutions to RefSet
   (remove the worst b1 sols. in Refset) */
for(i=b1+1;i<=pb->rs->b;i++)
{
  a=SSMax_dist_index(pb,min_dist);
  index = pb->rs->order[i];

  for(j=1;j<=pb->nvar;j++)
    pb->rs->sol[index][j] = pb->p->sol[a][j];
  pb->rs->ObjVal[index] = pb->p->ObjVal[a];
  pb->rs->iter[index] = pb->CurrentIter;

  SSUpdate_distances(pb,min_dist,index);
}

/* Reorder RefSet */
rs_order = SSOrder(pb->rs->ObjVal,pb->rs->b,pb->Opt);
for(i=1;i<=pb->rs->b;i++)
  pb->rs->order[i] = rs_order[i];

pb->rs->NewSolutions = 1;

free(min_dist+1);free(rs_order+1);
}
```

4.   SUBSET GENERATION METHOD

This method consists of generating subsets of reference solutions to be subjected to the Combination Method. The general scatter search framework considers the generation of subsets of size 2, 3 and 4. (As indicated in Chapter 1, subset sizes are denoted by s_{size}). In addition, the framework considers subset sizes larger than 4, by choosing the best i elements of *RefSet* and varying i from 5 to b. We will discuss these advanced designs in Chapter 5, however, for the purpose of our tutorial we limit our scope to

$s_{size} = 2$.   That is, we consider only subsets consisting of all pairwise combinations of the solutions in *RefSet*.

There are a maximum of $(b^2-b)/2$ subsets of this type, which in the case of our example amount to a total of $(10^2-10)/2 = 45$. The maximum number of subsets occurs when all the solutions in the reference set are new and have not been combined. For example, this happens at the beginning of the search when the reference set is constructed using the Diversification Generation Method. In subsequent iterations the maximum may not be reached because the Subset Generation Method discards pairs of reference solutions that have already been combined in previous iterations.

4.1   Computer Code

SSCombine_RefSet (Function 2-7) implements the Subset Generation Method and then calls the Combination Method. The procedure considers all pairs of solutions in the *Refset* that contain at least one new solution. A new solution is one that has not been combined in the past and therefore the procedure compares the iteration number in which a solution entered the reference set pb->rs->iter[i] with the current iteration number pb->CurrentIter. This means that the procedure does not allow for the same two solutions to be subjected to the Combination Method more than once. Once a new subset has been identified, the Combination Method (SSCombine) is called. The Combination Method generates three solutions every time is called and these solutions are stored in pool. The value of pool_size is also updated in Function 2-7.

Function 2-7. SSCombine_RefSet — File: SSRefSet1.c

```
void SSCombine_RefSet(SS *pb)
{
    int i,j,a,s;
    double **newsols;

    newsols = SSDouble_matrix(3,pb->nvar);

    /* Combine elements in RefSet */
    for(i=1;i<pb->rs->b;i++)
    for(j=i+1;j<=pb->rs->b;j++)
    {
      /* Combine solutions that have not been previously
         combined (3 new trial solutions for each pair)   */
      if(pb->rs->iter[i] == pb->CurrentIter ||
         pb->rs->iter[j] == pb->CurrentIter)
```

```
    {
        SSCombine(pb,pb->rs->sol[i],pb->rs->sol[j],newsols);
        for(a=1;a<=3;a++)  {
          pb->pool_size++;
          for(s=1;s<=pb->nvar;s++)
             pb->pool[pb->pool_size][s]=newsols[a][s];
    } } }
    SSFree_double_matrix(newsols,3);
}
```

5.   COMBINATION METHOD

This method uses the subsets generated with the Subset Generation Method to combine the solutions in each subset with the purpose of creating new trial solutions. The Combination Method is a problem-specific mechanism, because it is directly related to the solution representation. Depending on the specific form of the Combination Method, each subset can create one or more new solutions. For the purpose of this tutorial, we consider the following Combination Method for solutions that can be represented by bounded continuous variables.

The method consists of finding linear combinations of reference solutions. The number of solutions created from the linear combination of two reference solutions may depend on the quality of the solutions being combined. However, in this tutorial we simply create three trial solutions in every combination of two references solutions that we denote with x' and x'' :

C1. $x = x' - d$
C2. $x = x' + d$
C3. $x = x'' + d$

where $d = r\dfrac{x'' - x'}{2}$ and r is a random number in the range (0, 1). To illustrate the combination mechanism, consider the combination of the first two solutions in *RefSet* (i.e., solutions 35 and 46 in Table 2-4). Table 2-7 shows an instance of the three new trial solutions generated from combining these two high quality reference solutions. Solution 111 in Table 2-7 is constructed using the shown r-value and the combination equation C1 above. Similarly, solutions 112 and 113 are constructed using combinations C2 and C3, respectively.

Table 2-7. New trial solutions generated from the combination of solutions 35 and 46

Solution	r	x_1	x_2	x_3	x_4	$f(x)$
111	0.4498	-0.3091	-0.2295	2.2834	2.3484	753.72
112	0.4416	0.2204	0.3143	0.4319	1.2610	113.78
113	0.8997	1.3977	1.5458	-1.6256	-0.0350	671.64

Solutions generated with the Combination Method are subjected to the Improvement Method before they are considered for membership in the reference set. After applying the Improvement Method to the solutions in Table 2-7, the solutions are transformed to those shown in Table 2-8.

Table 2-8. Improved trial solutions

Solution	x_1	x_2	x_3	x_4	$f(x)$
111	-0.7384	0.5051	1.1794	1.3569	3.5823
112	-0.6192	0.3821	1.2149	1.4348	3.2672
113	1.2866	1.6275	0.3988	0.1248	1.4655

The best solution in Table 2-8 is C3 with an objective function value of 1.4655. Using the subsets that are generated with the SSCombine_RefSet function, more combinations are performed to create additional trial solutions. The search continues in a loop that consists of applying the Combination Method followed by the Improvement Method and the Reference Update Method. This loop terminates when the reference set does not change and all the subsets have already been subjected to the Combination Method. At this point, the Diversification Generation Method is used to rebuild half of the *RefSet* and the search continues. The complete procedure is outlined in Section 6 of this chapter.

5.1   Computer Code

The SSCombine function (see Function 2-8) produces three solutions by applying the linear combination equations C1, C2 and C3. The midpoint between solutions sol[i] and sol[j] is calculated and stored in the d array. This array is then used to construct new trial solutions employing a controlled randomization provided by the value of the r variable. In this implementation, we keep the random number r fixed for each solution type. Therefore, three random numbers are used in order to generate the desired number of solutions from the pair of reference solutions. The SSRandNum() function is a hardware independent number generator that is included in the SSTools1.c file. The function returns a pseudo-random number between zero and one.

Function 2-8. SSCombine — File: SSRefSet1.c

```
void SSCombine(SS *pb,double sol1[],double sol2[],double
**newsols)
{
    int j;
    double *d, r;
    d=SSDouble_array(pb->nvar);
    for(j=1;j<=pb->nvar;j++)
      d[j] = (sol2[j]-sol1[j])/2;

    /* Generate a C1 solution */
    r=SSRandNum();
    for(j=1;j<=pb->nvar;j++) {
      newsols[1][j] = sol1[j] - r*d[j];
      if(newsols[1][j]>pb->high[j]) newsols[1][j]=pb->high[j];
      if(newsols[1][j]<pb->low[j])  newsols[1][j]=pb->low[j];
    }
    /* Generate a C2 solution */
    r=SSRandNum();
    for(j=1;j<=pb->nvar;j++) {
      newsols[2][j] = sol1[j] + r*d[j];
      if(newsols[2][j]>pb->high[j]) newsols[2][j]=pb->high[j];
      if(newsols[2][j]<pb->low[j])  newsols[2][j]=pb->low[j];
    }
    /* Generate a C3 solution */
    r=SSRandNum();
    for(j=1;j<=pb->nvar;j++) {
      newsols[3][j] = sol2[j] + r*d[j];
      if(newsols[3][j]>pb->high[j]) newsols[3][j]=pb->high[j];
      if(newsols[3][j]<pb->low[j])  newsols[3][j]=pb->low[j];
    }
    free(d+1);
}
```

6.   OVERALL PROCEDURE

In the previous sections, we illustrated the operations performed within each of the methods in a scatter search implementation for unconstrained nonlinear optimization problems. We now finish our first tutorial with an overall view of the procedure. The outline (or pseudo-code) that summarizes the overall procedure uses the following parameters:

PSize	the size of the set P of diverse solutions generated by the Diversification Generation Method
b	the size of the reference set
b_1	the initial number of high-quality in the reference set
b_2	the initial number of diverse solutions in the reference set
MaxIter	maximum number of iterations

and the following variables:

Iter	counts the number of iterations
NewElements	a Boolean variable to indicate whether or not the updated reference set contains new solutions
Subsets	a list of the subsets generated with the Subset Generation Method

The procedure consists of the steps in the outline of Figure 2-1, where P denotes the set of solutions generated with the Diversification Generation Method, *RefSet* is the set of solutions in the reference set and *Pool* is the set of trial solutions constructed with the Combination and Improvement Methods. The procedure starts with the generation of *PSize* distinct solutions. These solutions are originally generated to be diverse and subsequently improved by the application of the Improvement Method (step 1). The set P of *PSize* solutions is ordered in step 2 to facilitate the task of creating the reference set. The first b_1 solutions in P are added to *RefSet* in this step.

After the initialization steps, the search consists of three main loops: 1) a "for-loop" that controls the maximum number of iterations, 2) a "while-loop" that monitors the presence of new elements in the reference set, and 3) a "while-loop" that controls the examination of all the subsets with at least one new element. In step 3, b_2 diverse solutions are added to *RefSet* to have a complete reference set with b solutions. The subsets with at least one new element are generated and added to *Subsets* in step 4. Also, the Boolean variable *NewElements* is turned into FALSE before the subsets are examined and the reference set is updated, since it is not known whether a new solution will enter the reference set in the current examination of the subsets. The subsets are used to create new trial solutions, which in turn are improved before being added to *Pool*. This is done in steps 5-8. The reference set is updated in step 9. If at least one of the improved solutions added to *Pool* in step 8 is different than the current reference solutions and is added to the new reference set in step 9 then the *NewElements* indicator is switched to TRUE in step 10.

1. Start with $P = \emptyset$. Use the Diversification Generation Method to construct a solution x. Apply the Improvement Method to x to obtain the improved solution x^*. If $x^* \notin P$ then, add x^* to P (i.e., $P = P \cup x^*$), otherwise discard x^*. Repeat this step until $|P| = PSize$.
2. Order the solutions in P according to their objective function value (where the best overall solution is first in the list), add the first b_1 solutions to *RefSet* and delete those solutions from P.

for (*Iter* = 1 **to** *MaxIter*)

    3. For each solution x in *P-RefSet* and y in *RefSet*, calculate the distance $d(x,y)$. Select the solution x' that maximizes $d_{min}(x)$, where

$$d_{min}(x) = \min_{y \in RefSet} \{d(x, y)\}.$$ Add x' to *RefSet* and delete it from P.

    Repeat this step b_2 times. Make *NewElements* = TRUE.

    while (*NewElements*) **do**

        4. Generate *Subsets* with the Subset Generation Method. This method generates all subsets of size 2; skipping subsets for which both elements have not changed from previous iterations. Make *NewElements* = FALSE.

        while (*Subsets* ≠ ∅) **do**

            5. Select the next subset s in *Subsets*.

            6. Apply the Combination Method to s to obtain three trial solutions x.

            7. Apply the Improvement Method to the trial solutions x, to obtain the improved solutions x^*.

            8. Add the improved solutions x^* to *Pool* and delete s from *Subsets*

        end while

        9. Update the reference set by selecting the best b solutions in *RefSet* $\cup$ *Pool*.

        if (updated reference set has at least one new solution) **then**

            10. Make *NewElements* = TRUE

    end while

    if (*Iter* < *MaxIter*) **then**

        11. Delete the last (worst) b_2 solutions from *RefSet*. Build a new set P using the Diversification Generation Method.

    end if

end for

Figure 2-1. Scatter search outline for unconstrained nonlinear optimization

If *NewElements* remains false, then no more trial solutions are generated from the current reference set. If at least one more iteration is available, then step 11 is performed. This step deletes the b_2 worst solutions from *RefSet* and rebuilds P. The new P is used to add b_2 solutions to *RefSet* when control goes back to step 3.

The outline in Figure 2-1 is a somewhat expanded and more explicit version of the basic outline in Figure 1-3. The outline in Figure 2-1 is more explicit because it specifies that:

–   solution dissimilarity is measured with Euclidean distances;
–   only subsets of size two are generated;
–   the reference set is initially built and later rebuilt with b_1 high-quality solutions and b_2 diverse solutions;
–   the procedure is performed *MaxIter* times;
–   the best solutions in *RefSet* are used to initialize *P*;

The structure in this outline is the same that we employ for the tutorials in Chapters 3 and 4. Hence, it is a good idea to take additional time to fully understand the login in the pseudo-code of Figure 2-1.

6.1   Computer Code

The computer codes for this tutorial can be found in the Chapter2 folder of the compact disc accompanying this book. Function 2-9 shows the main function of the scatter search code. It starts with the definition of the problem size `nvar`, the search parameters `b`, `PSize`, `MaxIter` and the iteration counter `Iter`. The search parameter values that appear in Function 2-9 are those recommended in the chapter. Although these values should yield reasonable results, we recommend that the reader experiments with different parameter values to investigate the effect that changing these values has on the quality of the best solution found.

The `SSProblem_Definition` function sets up the optimization problem. The function triggers the allocation of the required memory, including the reference set, the population *P,* and the auxiliary data structures in the problem. This function also specifies lower and upper bounds, l_i and u_i, for each decision variable x_i. The function returns the pointer `pb` to a memory structure that contains all the data associated with the problem to be solved.

The *P* set and the *RefSet* are created with calls to `SSCreate_P` and `SSCreate_RefSet`, respectively. The search takes place within the for-loop for the specified `MaxIter` iterations. At each iteration, if there are new solutions in the *RefSet* (i.e., when `pb->rs->NewSolutions` is true), they are combined and the *RefSet* is updated with a call to `SSUpdate_RefSet`; otherwise, it is partially rebuilt with a call to `SSRebuild_RefSet`. A typical run of this scatter search implementation executes the `SSUpdate_RefSet` function several times until the Combination Method is not able to produce a solution of a quality that warrants becoming a member of *RefSet*. In this case, the `SSRebuild_RefSet` function is executed. When the search is over,

the best solution found is printed and the allocated memory is free with a call
to SSFree_DataStructures, which deletes the previously defined problem
from memory.

Function 2-9. main — File: SSMain1.c

```c
#include "SS1.h"
int main(void)
{
    SS *pb;                      /* Pointer to problem data         */
    int nvar    =    4;          /* Number of variables             */
    int b       =   10;          /* Size of the reference set       */
    int PSize   =  100;          /* Size of P                       */
    int Iter    =    1;          /* Current iteration               */
    int MaxIter =  100;          /* Maximum number of iterations    */
    double  *sol, value;
    int i;

    pb = SSProblem_Definition(nvar,b,PSize);
    SSCreate_P(pb);
    SSCreate_RefSet(pb);

    for(Iter=1; Iter<MaxIter; Iter++)   {
      if(pb->rs->NewSolutions)
        SSUpdate_RefSet(pb);
      else
        SSRebuild_RefSet(pb);
    }

    /* Print Best Solution Found */
    sol = SSDouble_array(pb->nvar);
    SSBestSol(pb,sol,&value);
    printf("\n\nBest Solution Found:\nValue = %f",value);
    for(i=1;i<=nvar;i++) printf("\nx[%d]   = %f",i,sol[i]);
    free(sol+1);
    SSFree_DataStructures(pb);
    return 0;
}
```

The objective function is evaluated with a user-defined function, which
in this case we have named sol_value (Function 2-10). The single
argument of this function is the double vector sol and returns a single value.
Since this function is called from several points in the tutorial code, it is

necessary to keep the name of the function unchanged even if the equations within the function are changed. To solve a different problem than the one used in this tutorial, it suffices to change the calculations in Function 2-10, the variable bounds specified in SSProblem_Definition and the value of nvar in the main function.

Function 2-10. sol_value — File: SSMain1.c

```
double sol_value(double sol[])
{
    double value=0;

    value  = 100*pow(sol[2]-pow(sol[1],2),2)+pow(1-sol[1],2);
    value += 90 *pow(sol[4]-pow(sol[3],2),2)+pow(1-sol[3],2);
    value += 10.1*(pow(sol[2]-1,2)+pow(sol[4]-1,2));
    value += 19.8*(sol[2]-1)*(sol[4]-1);

    return value;
}
```

7.   SUMMARY OF C FUNCTIONS

Functions in file **SSImprove1.c**

```
void    SSImprove_solution(SS *pb, double sol[],double *value)
double  SSMove(SS *pb,double *worst_point,double *worst_value,
        double *psum, double factor)
void    SS_Simplex(SS *pb,double **simplex,double *values,int
        max_eval)
```

Functions in file **SSMemory1.c**

```
double *SSDouble_array(int size)
double **SSDouble_matrix(int nrows,int ncolumns)
void    SSFree_DataStructures(SS *pb)
void    SSFree_double_matrix(double **matrix,int nrows)
void    SSFree_int_matrix(int **matrix,int nrows)
int    *SSInt_array(int size)
int    **SSInt_matrix(int nrows,int ncolumns)
SS     *SSProblem_Definition(int nvar,int b,int PSize)
```

Functions in file **SSP1.c**

```
void    SSCreate_P(SS *pb)
void    SSGenerate_Sol(SS *pb,double sol[])
void    SSPrint_P(SS *pb)
```

Functions in file **SSRefSet1.c**

```
void    SSCombine(SS *pb,double sol1[],double sol2[],double
        **newsols)
void    SSCombine_RefSet(SS *pb)
void    SSCreate_RefSet(SS *pb)
void    SSPrint_RefSet(SS *pb)
void    SSRebuild_RefSet(SS *pb)
void    SSTryAdd_RefSet(SS *pb,double sol[],double value)
void    SSUpdate_RefSet(SS *pb)
```

Functions in file **SSTools1.c**

```
void    SSAbort(char texto[])
void    SSBestSol(SS *pb,double sol[],double *value)
double  SSDist_RefSet(SS *pb,int num,double sol[])
int     SSEqualSol(double sol1[],double sol2[],int dim)
int     SSIsInRefSet(SS *pb,double sol[])
int     SSMax_dist_index(SS *pb,double dist[])
int     *SSOrder(double weights[],int num,int type)
float   SSRandNum(void)
void    SSUpdate_distances(SS *pb,double min_dist[],int
        rs_index)
```

Chapter 3

TUTORIAL
0-1 Knapsack Problems

Y es que no hay nada mejor que formular, ... describiendo una trayectoria más.

Antonio Vega, Una Décima de Segundo (1985)

In the previous chapter, we described a basic implementation of scatter search in the context of nonlinear optimization. Our goal was to describe all five scatter search methods using straightforward strategies while developing a solution procedure capable of yielding high quality solutions. In this chapter, we will introduce some new strategies without excessively complicating the implementation. There are two main differences between the problem addressed in this tutorial and the problem addressed in the tutorial of the previous chapter:

– Binary variables are used to represent solutions
– The problem is constrained

Knapsack problems are well known in the operations research literature. The problem consists of choosing, from a set of items, the subset that maximizes the value of the objective function subject to a capacity constraint. Mathematically, the problem can be expressed as follows:

$$\text{Maximize} \quad \sum_i c_i x_i$$

$$\text{Subject to} \quad \sum_i a_i x_i \leq b$$

$$x_i \in \{0,1\} \qquad \forall i$$

This problem has many practical applications and the name is due to the situation that a hiker would face while selecting items to be carried in his/her backpack while keeping the total weight under a desired limit. Each item i is considered to have a certain utility value (c_i) and a known weight (a_i). Since the total weight of all the items under consideration exceeds the desired weight limit (b), the hiker must select the subset of items that maximizes the total utility. For a comprehensive examination of the knapsack and other related problems (including computer code for exact and heuristic procedures), we refer the reader to Martello and Toth (1989).

For the purpose of illustrating the scatter search mechanisms introduced in this chapter, let us consider the following instance of a 0-1 knapsack problem, where the coefficients in the objective function are profit values and the coefficients in the constraint are weights:

Maximize
$$11x_1 + 10x_2 + 9x_3 + 12x_4 + 10x_5 + 6x_6 + 7x_7 + 5x_8 + 3x_9 + 8x_{10}$$

Subject to
$$33x_1 + 27x_2 + 16x_3 + 14x_4 + 29x_5 + 30x_6 + 31x_7 + 33x_8 + 14x_9 + 18x_{10} \leq 100$$
$$x_i \in \{0, 1\} \text{ for } i = 1, ..., 10.$$

Following the same structure as in Chapter 2, we describe each of the methods that are needed in a scatter search implementation designed to find high-quality solutions to 0-1 knapsack problems and use the example problem to illustrate how the mechanisms work. The specific form of the methods suggested in this chapter is not unique. That is, there are multiple ways in which scatter search can be applied to the solution of 0-1 knapsack problems and our goal is not to explore all possible design choices but rather concentrate in one and illustrate how scatter search can deal with constrained binary problems.

1.   DIVERSIFICATION GENERATION METHOD

In the scatter search methodology, Diversification Generation Methods can be totally deterministic or partially random. That is, diversification can be implemented using systematic procedures or controlled randomization. Contrary to the philosophy of genetic algorithms, scatter search does not advocate the use of total randomization as a mechanism to achieve diversification. In the tutorial of Chapter 2, we introduced a Diversification Generation Method that was based on a randomization scheme that was controlled with a frequency memory mechanism. In this chapter, we illustrate the implementation of a systematic method for generating diverse solutions.

Glover (1998) describes the following systematic (i.e., non-random) procedures for generating diverse 0-1 vectors. We let x denote an n-vector, each of whose components x_i receives the value 0 or 1. Based on this definition, we first consider a diversification generation method that takes such a vector x as its seed solution, and generates a collection of solutions associated with an integer $h = 2, 3,..., h_{max}$, where $h_{max} \leq n$ - 1. Although it is recommended for h_{max} to be less than or equal to $n/5$, when a large number of diverse solutions is needed, the value of h_{max} must approach n - 1.

We generate two types of solutions, x' and x'', for each value of h, by the following rule:

Type 1 Solution: Let $x'_{1+kh} = 1 - x_{1+kh}$ for $k = 0, 1, 2, 3, ..., \lfloor n/h \rfloor$, where $\lfloor k \rfloor$ is the largest integer satisfying $k \leq n/h$. All other components of x' are equal to x (i.e., the seed solution).

Type 2 Solution: Let x'' be the complement of x'.

To illustrate, for the seed $x = (0, 0, ..., 0)$, the values $h = 2, ...$ 5 yield the solutions in Table 3-1. The first row corresponds to the seed and its complement and the remaining rows to the corresponding h values. The progression in Table 3-1 suggests the reason for preferring $h_{max} \leq n/5$. As h becomes larger, the solutions x' for two consecutive values of h differ from each other proportionally less than when h is smaller. An option to avoid diminishing diversity is to allow h to grow by increasing steps for larger values of h.

Table 3-1. Diverse 0-1 vectors

h	x'	x''
seed	(0,0,0,0,0,0,0,0,0,0)	(1,1,1,1,1,1,1,1,1,1)
2	(1,0,1,0,1,0,1,0,1,0)	(0,1,0,1,0,1,0,1,0,1)
3	(1,0,0,1,0,0,1,0,0,1)	(0,1,1,0,1,1,0,1,1,0)
4	(1,0,0,0,1,0,0,0,1,0)	(0,1,1,1,0,1,1,1,0,1)
5	(1,0,0,0,0,1,0,0,0,0)	(0,1,1,1,1,0,1,1,1,1)

The preceding design can be extended to generate additional solutions as follows. For values of $h \geq 2$ to $n-1$, the solution vector is shifted so that the index 1 is instead represented as a variable index q, which can take the values 1, 2, 3, ..., h. Continuing the illustration for the seed $x = (0, 0, ..., 0)$, suppose $h = 3$. Then, in addition to $x' = (1,0,0,1,0,0,1 ...)$, the method also generates the solutions given by $x' = (0,1,0,0,1,0,0,1, ...)$ and $x' = (0,0,1,0,0,1,0,0,1, ...)$, as q takes on the values 2 and 3. Specifically, for each value of $q = 1, 2, ..., h$, the following solutions are generated:

Type 1′ Solution: Let $x'_{q+kh} = 1 - x_{q+kh}$ for $k = 0, 1, 2, 3, ..., \lfloor (n-q)/h \rfloor$.
All other components of x' are equal to x (i.e., the seed solution).

Type 2′ Solution: Let x'' be the complement of x'.

This Diversification Generation Method is not capable of generating a large number of solutions. For example, in our illustrative problem that consists of 10 variables, the maximum number of solutions generated with this deterministic method is approximately 90 (without counting duplications). In order to obtain a larger number of solutions, the generator can be seeded employing different solutions. A simple deterministic implementation is to start with $x = (0, 0, ..., 0)$ and after generating all possible solutions from this seed then the generator is reinitialized using the last solution constructed in the previous run. This procedure would also generate duplications and therefore a checkup mechanism is needed to discard those solutions that are generated more than once.

Using the data in our example problem, the Diversification Generation Method is executed at the beginning of the search to generate a set P of 30 diverse solutions. Table 3-2 shows only the first ten. It should be noted that some of these solutions are infeasible and therefore their objective function value is larger than the optimum objective function value of 44.

Table 3-2. Diverse solutions

Sol.	x_1	x_2	x_3	x_4	x_5	x_6	x_7	x_8	x_9	x_{10}	$f(x)$
1	0	0	0	0	0	0	0	0	0	0	0
2	1	1	1	1	1	1	1	1	1	1	81
3	0	1	0	1	0	1	0	1	0	1	41
4	1	0	1	0	1	0	1	0	1	0	40
5	0	1	1	0	1	1	0	1	1	0	43
6	1	0	0	1	0	0	1	0	0	1	38
7	1	0	1	1	0	1	1	0	1	1	56
8	0	1	0	0	1	0	0	1	0	0	25
9	1	1	0	1	1	0	1	1	0	1	63
10	0	0	1	0	0	1	0	0	1	0	18

1.1   Computer Code

The `SSGenerate_Sols` function (see Function 3-1) in file SSP2.c of the Chapter3 folder implements the Diversification Generator Method, and generates a set of solutions each time it is called. These solutions are stored in the DivSol matrix which is part of the SS data structure. The DivSol matrix is allocated in SSProblem_Definition with a size of NdivSol rows and nvar columns. The SSTryAddSolDivSet function adds the current solution and its complement to the matrix if they do not duplicate any previous solution. The diversification generator is executed twice in the context of our illustrative example.

Function 3-1. SSGenerate_Sols — File: SSP2.c

```
void SSGenerate_Sols(SS *pb)
{
    int MaxSol,max_h,max_q,max_k;
    int h,q,k,iter=0,*sol,*seed;

    sol  = SSInt_array(pb->nvar);
    seed = SSInt_array(pb->nvar);

    MaxSol   = pb->NDivSol;
    max_h    = pb->nvar-1;
    pb->CDivSol = 0;

    for(k=1;k<=pb->nvar;k++) sol[k]=0;
    SSTryAddSolDivSet(pb,sol);

    while(iter++<2  &&  pb->CDivSol<=MaxSol)
    {
      /* Seed with the last solution */
      for(k=1;k<=pb->nvar;k++)
        seed[k]=sol[k];

      h=2;
      while(h<=max_h && pb->CDivSol<=MaxSol)
      {
        q=1;max_q=h;
        while(q<=max_q && pb->CDivSol<MaxSol-1)
        {
          for(k=1;k<=pb->nvar;k++)
            sol[k]=seed[k];

          max_k=(int)(pb->nvar-q)/h;
          for(k=0;k<=max_k;k++)
            sol[q+k*h]=1-seed[q+k*h];
```

```
        /* Add solution and its complement */
        SSTryAddSolDivSet(pb,sol);
        q++;
      }
      h++;
    }
  }
  free(sol+1);free(seed+1);
}
```

The SScreate_P function (see Function 3-2) in file SSP2.c builds a population of PSize elements.   Function 3-2 begins with a call to SSGenerate_Sol and then extracts PSize different solutions from the DivSol matrix.   Each element is submitted to SSImprove_solution for feasibility, if necessary, and improvement, if possible.   Then, the SSEqualSol function checks whether the newly extracted and improved solution is different from all previously added to *P*. The function ends when PSize different improved solutions have been added to *P*.   If the diversification generator is not able to create PSize different solutions or too many of them converge to the same local optima after they are subjected to the Improvement Method, a call to the SSAbort function prints the message "Reduce the PSize value" and the program execution finishes.

Function 3-2. SSCreate_P — File: SSP2.c

```
void SSCreate_P(SS *pb)
{
    int obj_val,*sol,currentPSize,j,equal,DivSolCount;
    sol = SSint_array(pb->nvar);
    SSGenerate_Sols(pb);
    if(pb->CDivSol < pb->p->PSize)
      SSAbort("Reduce the PSize value");

    DivSolCount=currentPSize=1;
    while(currentPSize<=pb->p->PSize &&
          DivSolCount<=pb->CDivSol)
    {
      obj_val = sol_value(pb,pb->DivSol[DivSolCount]);
      SSImprove_solution(pb,pb->DivSol[DivSolCount],&obj_val);

      /* Check if sol is a new one */
      j=1;equal=0;
      while(j<currentPSize && !equal)
```

```
        equal=SSEqualSol(pb->DivSol[DivSolCount],
                        pb->p->sol[j++],pb->nvar);

    /* Add solution to set P */
    if(!equal) {
        for(j=1;j<=pb->nvar;j++)
    pb->p->sol[currentPSize][j]=pb->DivSol[DivSolCount][j];
        pb->p->ObjVal[currentPSize++] = obj_val;
    }
    DivSolCount++;
    }
    if(currentPSize<pb->p->PSize)
        SSAbort("Reduce the PSize value");

    free(sol+1);
}
```

2.   IMPROVEMENT METHOD

Since the solutions generated with the Diversification Generation Method are not guaranteed to be feasible, the Improvement Method should be capable of handling starting solutions that are either feasible or infeasible. To illustrate the development of such a method, we consider a straightforward two-phase procedure for knapsack problems that operates as follows:

Feasibility Phase

If the trial solution is *infeasible*, then the method attempts to improve it by first finding a feasible solution. A feasible solution is found by changing variable values from one to zero until the constraint is no longer violated. The variables are considered in increasing order of their profit-to-weight ratio, starting with the one with the smallest ratio. Once the solution becomes feasible, the Improvement Phase is applied.

Improvement Phase

If the trial solution is *feasible*, then the method attempts to improve it by changing variable values from zero to one. This is done in a greedy fashion, so that the first variable to be considered is the one with the largest profit-to-weight ratio. (For example, x_4 has the largest ratio value of $12/14 = 0.857$ in our problem instance.) The procedure stops when no more variables can be given a value of one without violating the constraint.

Table 3-3 shows the solutions obtained by the application of the Feasibility Phase to the trial solutions in Table 3-2 generated with the Diversification Generation Method. Since solutions 1, 6, 8 and 10 are already feasible they remain unchanged from Table 3-2 to Table 3-3. All other solutions experience deterioration in their objective function value when turning from infeasible to feasible.

Table 3-3. Feasible and improved solutions

Solution	Feasible Solution	Objective Value	Improved Solution	Objective Value
1	(0,0,0,0,0,0,0,0,0,0)	0	(0,1,1,1,0,0,0,0,0,1)	39
2	(0,1,1,1,0,0,0,0,0,1)	39	(0,1,1,1,0,0,0,0,0,1)	39
3	(0,1,0,1,0,1,0,0,0,1)	36	(0,1,0,1,0,1,0,0,0,1)	36
4	(1,0,1,0,1,0,0,0,0,0)	30	(1,0,1,1,1,0,0,0,0,0)	42
5	(0,1,1,0,1,0,0,0,1,0)	32	(0,1,1,1,1,0,0,0,1,0)	44
6	(1,0,0,1,0,0,1,0,0,1)	38	(1,0,0,1,0,0,1,0,0,1)	38
7	(1,0,1,1,0,0,0,0,0,1)	40	(1,0,1,1,0,0,0,0,0,1)	40
8	(0,1,0,0,1,0,0,1,0,0)	25	(0,1,0,0,1,0,0,1,0,0)	25
9	(0,1,0,1,1,0,0,0,0,1)	40	(0,1,0,1,1,0,0,0,0,1)	40
10	(0,0,1,0,0,1,0,0,1,0)	18	(0,0,1,1,0,1,0,0,1,1)	38

The Improved Solutions in Table 3-3 are the result of applying the Improvement Phase of our Improvement Method. Note that the feasible solutions labeled 1 and 2 in Table 3-3 became the same improved solution with an objective function value of 39 after the application of the Improvement Phase. In these cases, which are not all that unusual, the Diversification Generation Method can be applied multiple times until a desired number of improved solutions that are different from each other are found. Also note that the Improvement Method was able to find the optimal solution starting from the infeasible trial solution 5. Table 3-4 shows the moves that the Improvement Method performed to obtain the optimal solution starting from this infeasible trial solution. At iteration 3 the solution is made feasible and matches the one shown in Table 3-3.

Table 3-4. Iterations of the Improvement Method

Iteration	Current Solution	Objective value	Total Weight	Candidate Moves (profit/weight)	Selected Move
1	(0,1,1,0,1,1,0,1,1,0)	43	149	$x_2 = 0$ (0.370) $x_3 = 0$ (0.563) $x_5 = 0$ (0.345) $x_6 = 0$ (0.200) $x_8 = 0$ (0.152) $x_9 = 0$ (0.214)	$x_8 = 0$
2	(0,1,1,0,1,1,0,0,1,0)	38	116	$x_2 = 0$ (0.370) $x_3 = 0$ (0.563) $x_5 = 0$ (0.345) $x_6 = 0$ (0.200) $x_9 = 0$ (0.214)	$x_6 = 0$
3	(0,1,1,0,1,0,0,0,1,0)	32	86	$x_4 = 1$ (0.857)	$x_4 = 1$
4	(0,1,1,1,1,0,0,0,1,0)	44	100	None	

The iterations in Table 3-4 show that as long as the knapsack constraint is violated (i.e., the total weight of the current solution is larger than 100 in our example), the candidate moves consist of changing variable values from one to zero. In the first iteration, x_8 is selected because its profit-to-weight ratio is the smallest in the candidate list of moves. After x_6 is also set to zero in iteration 2 using the same criteria as before, the current solution becomes feasible in iteration 3. The procedure now looks for opportunities to improve upon the objective function value. The candidates are those variables that are currently set to zero (excluding x_6 and x_8) and whose weight is less than or equal to the current slack (i.e., 100-86 = 14). The only variable that meets such criteria is x_4, which also happens to have the largest profit-to-weight ratio of 0.857. When the value of x_4 is changed to 1, the optimal solution is found. No more moves are available in iteration 4 and the improvement process stops.

2.1   Computer Code

Function 3-3, named `SSImprove_solution`, implements both phases of the Improvement Method. `SSImprove_solution` has three arguments: 1) a pointer to the data structure that contains all the problem information, 2) an array with the solution to be improved, and 3) a pointer to the corresponding objective function value. The function first computes the profit-to-weight ratios and orders the variables according to these values with a call to `SSOrder_d`. Then, the Feasibility Phase is executed with a while-loop that checks, with a call to `SSIsFeasible`, whether the solution has become feasible after a variable value has been changed from one to zero. When a variable value is switched from one to zero, the objective function value is updated by subtracting the corresponding objective function coefficient from the current objective function value. The Improvement Phase is executed next with another while-loop that switches variable values from zero to one as long as the solution remains feasible. In this case, the objective function value is updated by adding the objective function coefficient of a variable whose value has been switched from zero to one. This phase searches for the variables that are currently set to zero and attempts a switch to a value of one. The phase terminates when no variable that is currently set to zero can be set to one without violating the knapsack constraint. When both phases are completed, sol contains the improved solution and value its corresponding objective function value.

Function 3-3. SSImprove_solution — File: SSImprove2.c

```
void SSImprove_solution(SS *pb,int sol[],int *value)
{
   int i,j,feasible,*order,lhs=0;
   double *ratio;

   /* Compute profit-to-weight ratio */
   ratio=SSDouble_array(pb->nvar);
   for(i=1;i<=pb->nvar;i++)
     ratio[i]=(double)pb->ObjCoeff[i]/(double)pb->ConstCoeff[i];
   order=SSOrder_d(ratio,pb->nvar,1);

   /* Feasibility Phase */
   j=pb->nvar;
   while(!SSIsFeasible(pb,sol))
   {
     if(sol[order[j]]==1)
     {
       sol[order[j]]=0;
       *value -= pb->ObjCoeff[order[j]];
     }
     j--;
   }

   /* Improvement Phase */
   j=feasible=1;
   for(i=1;i<=pb->nvar;i++)
     lhs += pb->ConstCoeff[i]*sol[i];

   while(feasible && j<=pb->nvar)
   {
```

```
/* Find first null value in sol */
while(sol[order[j]])
   j++;
if(lhs+pb->ConstCoeff[order[j]] <=
   pb->ConstCoeff[pb->nvar+1])
{
   sol[order[j]]=1;
   lhs += pb->ConstCoeff[order[j]];
   *value += pb->ObjCoeff[order[j]];
}
else feasible=0;
}

free(ratio+1);free(order+1);
}
```

The SSOrder_d and SSIsFeasible functions can be found in SSTools2.c while sol_value is in SSMain2.c. These files are in the Chapter3 directory of the accompanying disc. A summary of all the functions in this tutorial is provided at the end of the chapter.

3.   REFERENCE SET UPDATE METHOD

This method is used to create and maintain a set of reference solutions (*RefSet*). In this example, we also use a reference set of b solutions, which is initially populated with b_1 high-quality solutions and b_2 diverse solutions, with $b = b_1 + b_2$. For the purpose of this tutorial, we consider a reference set of size $b = 10$, where $b_1 = 5$ and $b_2 = 5$. According to these definitions, the high-quality solutions that are added to the initial reference set are solutions 4, 5, 18, 19 and 25 in P, which contains 30 non-duplicated solutions. Table 3-5 shows these solutions with their objective function values.

Table 3-5. High quality solutions in *RefSet*

Sol.	x_1	x_2	x_3	x_4	x_5	x_6	x_7	x_8	x_9	x_{10}	$f(x)$
4	0	1	1	1	1	0	0	0	1	0	44
5	1	0	1	1	0	0	0	0	1	1	43
18	1	0	1	1	1	0	0	0	0	0	42
19	0	1	1	1	0	0	0	0	1	1	42
25	1	1	1	1	0	0	0	0	0	0	42

In order to find diverse solutions to add to the reference set, it is necessary to define a diversity measure. We define the distance between two solutions as the sum of the absolute difference between its corresponding variable values. That is,

$$d(x, y) = \sum_i abs(x_i - y_i)$$

For example, the distance between the improved solution 1 and the improved solution 8 in Table 3-3 is calculated as follows:

```
(0,1,1,1,0,0,0,0,0,1)
(0,1,0,0,1,0,0,1,0,0)
(0+0+1+1+1+0+0+1+0+1) =  5
```

We then use this distance measure to select five solutions to complete the initial reference set. In particular, we look for a solution that is not currently in the reference set and that maximizes the minimum distance to all the solutions currently in the reference set. Table 3-4 shows the distance values from the first five solutions currently in P to each solution currently in the reference set.

Table 3-6. Distance from solutions in P to solutions in *RefSet*

Candidate Solution in P	Distance to solution					Minimum
	4	5	18	19	25	distance
1	3	3	4	1	2	1
2	5	5	6	3	4	3
3	7	3	4	5	4	3
6	5	1	2	3	2	1
7	4	8	5	6	5	4

The list of candidate solutions in Table 3-6 includes neither solution 4 nor solution 5 because these solutions are already in the reference set. The distance values show that in order to maximize the minimum distance, improved solution 7 should be added to the reference set. Although we only show 5 out of the 30 solutions in P, solution 7 in Table 3-6 still becomes a member of *RefSet* when all 30 solutions are considered. It is important to realize that the reference set does not consist of all the "best" solutions as measured by the objective function value only. Because we are interested in a balance between high quality solutions and diverse solutions, improved solution 1 with an objective function value of 39 was by-passed in favor of other solutions that add more diversity to the reference set. As shown in Table 3-6, the distance between solution 1 and solution 19 is only one unit, making this solution unattractive from the standpoint of diversification.

After the initial reference set has been built, the set is modified by replacing solutions using the static update method described in Section 3 of Chapter 2.

3.1   Computer Code

The SSCreate_RefSet function builds the reference set. Function 3-4 is essentially the same implementation as Function 2-4 of Chapter 2. The main difference is that the solutions and their objective function values in this tutorial are represented with variables of the int type, while in the previous tutorial they were stored in variables of the double type.

The procedure starts with a call to SSOrder_i to order the solutions in P (pb->p) according to the objective function values. Then the first b1 solutions in pb->p are added to the reference set pb->rs. The minimum distances from the first b1 solutions in pb->p to the other solutions in pb->p are calculated with a call to SSdist_RefSet. The most diverse b-b1 solutions are added to pb->rs with calls to SSmax_dist_index. Finally, the reference set is reordered and the Boolean variable NewSolutions and the iteration counter CurrentIter are both set to 1.

Function 3-4. SSCreate_RefSet — File: SSRefSet2.c

```
void SSCreate_RefSet(SS *pb)
{
    int b1,i,j,a,*p_order,*rs_order,*min_dist;

    /* Order solutions in P */
    p_order = SSOrder_i(pb->p->ObjVal,pb->p->PSize,pb->Opt);

    /* Add the b1 best solutions in P to RefSet */
    b1 = pb->rs->b / 2;
    for(i=1;i<=b1;i++)
    {
        for(j=1;j<=pb->nvar;j++)
            pb->rs->sol[i][j] = pb->p->sol[p_order[i]][j];
        pb->rs->ObjVal[i] = pb->p->ObjVal[p_order[i]];
        pb->rs->iter[i]   = 1;
    }

    /* Compute minimum distances from P to RefSet */
    min_dist = SSInt_array(pb->p->PSize);
    for(i=1;i<=pb->p->PSize;i++)
        min_dist[i]= SSdist_RefSet(pb,b1,pb->p->sol[i]);
```

```
/* Add the diverse b-b1 solutions to RefSet */
for(i=b1+1;i<=pb->rs->b;i++)
{
   a=SSMax_dist_index(pb,min_dist);
   for(j=1;j<=pb->nvar;j++)
      pb->rs->sol[i][j] = pb->p->sol[a][j];
   pb->rs->ObjVal[i] = pb->p->ObjVal[a];
   pb->rs->iter[i]   = 1;

   SSUpdate_distances(pb,min_dist,i);
}

/* Reorder RefSet */
rs_order = SSOrder_i(pb->rs->ObjVal,pb->rs->b,pb->Opt);
for(i=1;i<=pb->rs->b;i++)
   pb->rs->order[i] = rs_order[i];

pb->rs->NewSolutions = 1;
pb->CurrentIter      = 1;

free(rs_order+1);free(p_order+1);free(min_dist+1);
}
```

4.   SUBSET GENERATION METHOD

In this illustrative scatter search implementation we limit our scope to subsets of size 2 in the Subset Generation Method. Since the generation of subsets is context independent, the Subset Generation Method described in Section 4 of Chapter 2, which was designed for combining solution pairs, can be directly applied to the scatter search procedure developed in this chapter for 0-1 knapsack problems. Therefore, in order to avoid unnecessary duplication of material, we refer the reader to Section 4 of Chapter 2 for a detailed explanation of the Subset Generation Method for subset size equal to 2. For convenience, a copy of the code that implements the Subset Generation Method is included in the Chapter3 folder.

5. COMBINATION METHOD

This method uses the subsets generated with the Subset Generation Method to combine the elements in each subset with the purpose of creating new trial solutions. As mentioned before, the Combination Method is typically a problem-specific procedure. The implementation depends on the solution representation, which in the case of knapsack problems consists of a vector of 0-1 values (or also referred to as *binary string*). Combination Methods can be either systematic or probabilistic. In this tutorial, we illustrate the use of a probabilistic rule designed to generate a specified number of trial solutions from two reference solutions. A systematic Combination Method is described in Chapter 4 in the context of the linear ordering problem. The design can also be used to combine more than two solutions, as discussed in Chapter 5. Specifically, our Combination Method calculates a score for each variable, based on the objective function value of the two reference solutions being combined. The score for variable i that corresponds to the combination of reference solutions j and k is calculated with the following formula:

$$score(i) = \frac{ObjVal(j)x_i^j + ObjVal(k)x_i^k}{ObjVal(j) + ObjVal(k)}$$

Where $ObjVal(j)$ is the objective function value of solution j and x_i^j is the value of the ith variable in solution j. Then, the trial solution is constructed by using the score as the probability for setting each variable to one, i.e., $P(x_i = 1) = score(i)$. This can be implemented as follows:

$$x_i = \begin{cases} 1 & \text{if } r \leq score(i) \\ 0 & \text{if } r > score(i) \end{cases}$$

where r is a uniform random number such that $0 \leq r \leq 1$. Although in this tutorial a single value of r is used for all variables in a given combination step, other designs may include drawing one value of r for each variable in the same combination, as discussed in Chapter 5. A deterministic version of this Combination Method is obtained by fixing r to a certain value (e.g., $r = 0.5$) for all combination steps.

The combination mechanism may construct infeasible solutions. This does not represent a problem, since the Improvement Method is always applied to each trial solution created after the application of the Combination Method. Recall that the Improvement Method was designed to deal with either feasible or infeasible starting solutions.

To illustrate the use of the Combination Method, consider the combination of the two reference solutions labeled 4 and 5 in Table 3-5. The objective function values of these solutions are 44 and 43, i.e., $ObjVal(4) = 44$ and $ObjVal(5) = 43$. The score for each variable can be calculated as shown in Table 3-7.

Table 3-7. Combination of solutions 4 and 5

Sol.	x_1	x_2	x_3	x_4	x_5	x_6	x_7	x_8	x_9	x_{10}
4	0	1	1	1	1	0	0	0	1	0
5	1	0	1	1	0	0	0	0	1	1
Score	0.49	0.51	1	1	0.51	0	0	0	1	0.49

If we use $r = 0.5$ for all variables, the scores in the last row of Table 3-7 become the new solution $(0,1,1,1,1,0,0,0,1,0)$ with an objective value of 44 and a total weight of 100. This feasible solution cannot be improved with the application of the Improvement Method, since no variable values can be switched from zero to one without violating the constraint. The new trial solution is in fact the same as solution 4.

Table 3-8 shows the solutions in the reference set with their respective objective function values after the first iteration. That is, this is the reference set after it has been updated to consider the solutions initially in the set and the trial solutions obtained with the application of the Combination Method. The solutions in Table 3-8 have an average quality that is higher than the average quality of the solutions in the initial reference set. This is the effect of updating the reference set using a "highest quality" criterion, as done in this tutorial.

Table 3-8. RefSet at second iteration

x_1	x_2	x_3	x_4	x_5	x_6	x_7	x_8	x_9	x_{10}	$f(x)$
0	1	1	1	1	0	0	0	1	0	44
1	0	1	1	0	0	0	0	1	1	43
1	0	1	1	1	0	0	0	0	0	42
0	1	1	1	0	0	0	0	1	1	42
1	1	1	1	0	0	0	0	0	0	42
0	1	1	1	1	0	0	0	0	0	41
1	0	1	1	0	0	0	0	0	1	40
0	1	0	1	1	0	0	0	0	1	40
0	1	1	1	0	0	0	0	0	1	39

5.1   Computer Code

The `sscombine` function implements the variant of the Combination Method that generates new trial solutions with the combination of two reference solutions (Function 3-5). The score for each variable is calculated first and stored in the `score` array. The reference solutions being combined are `sol1` and `sol2` with objective function values given by `value1` and `value2`. A single random number `r` is generated for each new trial solution. The procedure generates two new trial solutions comparing the score values with the value of the random number. The new solutions are stored in the `newsols` matrix.

Function 3-5. SSCombine — File: SSRefSet2.c

```
void SSCombine(SS *pb, int sol1[], int sol2[], int value1,
            int value2,int **)
{
    int i,j;
    double r,*score;
    score=SSDouble_array(pb->nvar);

    /* Calculate scores */
    for(j=1;j<=pb->nvar;j++)
        score[j] = (double)(sol1[j]*value1+sol2[j]*value2)/
                (double)(value1+value2);

    /* Generate solutions */
    for(i=1;i<=2;i++)
    {
        r=SSRandNum();
        for(j=1;j<=pb->nvar;j++)
        {
            if(r<=score[j]) newsols[i][j] = 1;
            else            newsols[i][j] = 0;
        }
    }
    free(score+1);
}
```

6.   OVERALL PROCEDURE

The overall procedure for this tutorial is the same as the one outlined in Figure 2-1. That is, the 5 methods are integrated in the framework shown in Figure 2-1 with the difference that the Diversification Generation, Improvement and Combination methods are specific to 0-1 knapsack problems. The reference set manipulation in this tutorial is the same as the one used for the nonlinear optimization tutorial, and therefore, the Reference Set Update Method is the same for both procedures. Also, the Subset Generation Method is the same in both tutorials because both are limited to combinations of two reference solutions.

6.1   Computer Code

The computer code for the overall procedure follows the same logic as the one described in Section 6.1 of Chapter 2. In Function 3-6, we limit the size of *P* to 30 (i.e., PSize = 30) because the example problem we use for this tutorial is small and the application of the Improvement Method to every solution created with the Diversification Generation Method results in a fair amount of duplications. That is, feasible and infeasible trial solutions converge to the same local optimum when the Improvement Method is applied. Specifically, the more than 100 solutions that the Diversification Generation Method is capable of producing converge to 32 distinct local optima after the application of the Improvement Method.

The problem to be solved is defined with a call to SSProblem_Definition. Then, *P* and *RefSet* are created with sequential calls to SSCreate_P and SSCreate_RefSet. The search is performed with a for-loop that calls SSUpdate_RefSet and SSRebuild_RefSet. When the search is completed, the best solution is retrieved with a call to SSBestSol. Finally, all data structures and allocated memory are freed with a call to SSFree_DataStructures.

Function 3-6. Main — File: SSMain2.c

```
int main(void)
{
    SS *pb;                 /* Pointer to problem data structure  */
    int nvar     =  10;  /* Number of variables                */
    int b        =  10;  /* Size of reference set              */
    int PSize    =  30;  /* Size of P                          */
    int Iter     =   1;  /* Current iteration                  */
    int MaxIter  =  50;  /* Maximum number of iterations       */
    int *sol,value,i;

    pb = SSProblem_Definition(nvar,b,PSize);

    SSCreate_P(pb);
    SSCreate_RefSet(pb);
```

```
     for(Iter=1; Iter<MaxIter; Iter++)
     {
          if(pb->rs->NewSolutions)
            SSUpdate_RefSet(pb);
          else
            SSRebuild_RefSet(pb);
     }

     /* Print Best Solution Found */
     sol = SSint_array(pb->nvar);
     SSBestSol(pb,sol,&value);
     printf("\n\nBest Solution Found:\nValue = %d",value);
     for(i=1;i<=nvar;i++) printf("\nx[%d]   = %d",i,sol[i]);
     free(sol+1);

     SSFree_DataStructures(pb);
     return 0;
}
```

To solve a different problem than the one used to illustrate this scatter search implementation, it is necessary to change the input data in `SSSet_Coefficients`. Also, experiments may be conducted to test the effects of changing the search parameters that are set at the beginning of Function 3-6. Although these parameters are "hard-coded" in this main function to facilitate the description of the code, it may be more practical to modify the code to add these variables as arguments when performing extensive experimentation.

7.   SUMMARY OF C FUNCTIONS

Functions in file **SSImprove2.c**

```
void    SSImprove_solution(SS *pb, int sol[],int *value);
```

Functions in file **SSMemory2.c**

```
double  *SSDouble_array(int size);
double  **SSDouble_matrix(int nrows,int ncolumns);
void    SSFree_DataStructures(SS *pb);
void    SSFree_double_matrix(double **matrix,int nrows);
```

```
void    SSFree_int_matrix(int **matrix,int nrows);
int     *SSInt_array(int size);
int     **SSInt_matrix(int nrows,int ncolumns);
SS      *SSProblem_Definition(int nvar,int b,int PSize);
void    SSSet_Coefficients(SS *pb);
```

Functions in file **SSP2.c**

```
void    SSCreate_P(SS *pb);
void    SSGenerate_Sols(SS *pb);
void    SSPrint_P(SS *pb);
void    SSTryAddSolDivSet(SS *pb,int sol[]);
```

Functions in file **SSRefSet2.c**

```
void    SSCombine(SS *pb,int sol1[],int sol2[],int value1,int
        value2,int **newsols);
void    SSCombine_RefSet(SS *pb);
void    SSCreate_RefSet(SS *pb);
void    SSPrint_RefSet(SS *pb);
void    SSRebuild_RefSet(SS *pb);
void    SSTryAdd_RefSet(SS *pb,int sol[],int value);
void    SSUpdate_RefSet(SS *pb);
```

Functions in file **SSTools2.c**

```
void    SSAbort(char texto[]);
void    SSBestSol(SS *pb,int sol[],int *value);
int     SSDist_RefSet(SS *pb,int num,int sol[]);
int     SSEqualSol(int sol1[],int sol2[],int dim);
int     SSIsFeasible(SS *pb,int sol[]);
int     SSIsInDivSet(SS *pb,int sol[]);
int     SSIsInRefSet(SS *pb,int sol[]);
int     SSMax_dist_index(SS *pb,int dist[]);
int     *SSOrder_d(double weights[],int num,int type);
int     *SSOrder_i(int weights[],int num,int type);
float   SSRandNum(void);
void    SSUpdate_distances(SS *pb,int min_dist[],int rs_index);
```

Chapter 4

TUTORIAL
Linear Ordering Problem

Well, I don't think there is any question about it. It can only be attributable to human error.

HAL 9000 in 2001 Space Odyssey (1968)

This is our third and last tutorial chapter. The goal of this chapter is to illustrate the development of a scatter search procedure for a combinatorial optimization problem. We focus on the description of the context-dependent elements of scatter search. That is, while we provide details and computer code for the Diversification Generation, Improvement and Combination Methods, we only include a few comments regarding the Reference Set Update Method. The Subset Generation Method is identical to the one used in the first two tutorials and therefore we have left it out of this chapter to avoid being repetitive. The implementations described in Chapters 2 and 3 were developed for illustration purposes and well-known methods exist that might be more efficient than scatter search for the chosen problems (i.e., unconstrained nonlinear optimization and knapsack problems). However, the implementation in this chapter is based on a highly competitive procedure for solving the linear ordering problem. Additional interesting features of this tutorial are the use of (1) GRASP constructions for creating diverse solutions, (2) highly specialized Improvement Method and (3) a voting scheme to combine reference solutions.

1. THE LINEAR ORDERING PROBLEM

The linear ordering problem (or LOP) has generated a considerable amount of research interest over the years, as documented in Grotschel, et al. (1984) and Chanas and Kobylanski (1996). Because of its practical and theoretical relevance for a significant range of global optimization applications, we use this problem to illustrate the implementation of scatter search strategies and search mechanisms in the context of a combinatorial optimization problem.

The LOP consists of finding a permutation p of the columns (and rows) of a matrix of weights $E = \{\ e_{ij}\ \}_{m \times m}$, in order to maximize the sum of the weights in the triangle above the main diagonal. In mathematical terms, we seek to maximize:

$$C_E(p) = \sum_{i=1}^{m-1} \sum_{j=i+1}^{m} e_{p(i)p(j)}$$

where $p(i)$ is the index of the column (and row) in position i in the permutation. Note that in the LOP, the permutation p provides the ordering of both the columns and the rows. The equivalent problem in graphs consists of finding, in a complete weighted graph, an acyclic tournament with a maximal sum of arc weights (Reinelt, 1985).

In economics, the LOP is equivalent to the so-called *triangulation problem for input-output tables*, which can be described as follows. The economy of a region (generally a country) is divided into m sectors and an $m \times m$ input-output table E is constructed where the entry e_{ij} denotes the amount of deliveries (in monetary value) from sector i to sector j in a given year. The triangulation problem then consists of simultaneously permuting the rows and columns of E, to make the sum of the entries above the main diagonal as large as possible. An optimal solution then orders the sectors in such a way that the suppliers (i.e., sectors that tend to produce materials for other industries) come first, followed by the consumers (i.e., sectors that tend to be final-product industries that deliver their output mostly to end users). Instances of input-output tables from sectors in the European Economy can be found in the public-domain library LOLIB (1997).

Another reason for using the linear ordering problem in this tutorial is that several of the strategies associated with adapting scatter search for the LOP can be used for other problems for which the natural representation of a solution is a permutation. One of the best procedures for the solution of the LOP are due to Campos, et al. (2001) and Laguna, Martí and Campos (1999). The implementation described in this tutorial is based on Campos, et al. (2001). Also, we base the Improvement Method on the local search procedure designed by Laguna, Martí and Campos (1999).

2. DIVERSIFICATION GENERATION METHOD

The Diversification Generation Method is based on GRASP constructions. GRASP, *greedy randomized adaptive search procedure*, is a multi-start or iterative process, in which each iteration consists of two phases: construction and local search. The construction phase builds a feasible solution, whose neighborhood is explored until a local optimum is found after the application of the local search phase. The best local optimum is reported as the best overall solution found. Resende and Ribeiro (2001) present a comprehensive review of GRASP and an extensive survey of the GRASP literature can be found in Festa and Resende (2001).

Since our Diversification Generation Method is based on GRASP constructions, we provide a high-level description of this technique. Our description was adapted from Resende and Ribeiro (2001). At each iteration of the construction phase, GRASP maintains a set of candidate elements that can be feasibly added to the partial solution under construction. All candidate elements are evaluated according to a greedy function in order to select the next element to be added to the construction. The greedy function represents the marginal increase in the cost function from adding the element to the partial solution. The evaluation of the elements is used to create a restricted candidate list (RCL). RCL consists of the best elements, i.e. those with the smallest incremental cost. This is the greedy aspect of the method. The element to be added into the partial solution is randomly selected from those in the RCL. This is the probabilistic aspect of the heuristic. Once the selected element is added to the partial solution, the candidate list is updated and the incremental costs are recalculated. This is the adaptive aspect of the heuristic. This strategy is similar to the semi-greedy heuristic proposed by Hart and Shogan (1987), which is also a multi-start approach based on greedy randomized constructions, but without local search.

The solutions generated by a greedy randomized construction are not necessarily optimal, even with respect to simple neighborhoods. The local search phase usually improves upon the constructed solution. A local search algorithm works in an iterative fashion by successively replacing the current solution by a better solution in the neighborhood of the current solution. It terminates when no better solution is found in the neighborhood.

The effectiveness of a local search procedure depends on several aspects, such as the neighborhood structure, the neighborhood search techniques, the speed of evaluation of the objective function of neighbor solutions, and the initial solution. The construction phase plays a critical role with respect to providing high-quality starting solutions for the local search. Simple neighborhood structures based on swap and insert moves or add-delete moves are typically used. The neighborhood search may be implemented using either a best-improving or a first-improving strategy. In the case of best-improving strategies, all neighbors are examined and the best neighbor replaces the current solution. In the case of first-improving strategies, the search moves to the first neighbor whose cost function value is strictly less (by any margin) than that of the current solution.

A particularly appealing characteristic of GRASP is ease of implementation. Most implementations have a very small number of parameters and therefore developers can focus on implementation efficiency to speedup data manipulation and increase the number of iterations that can be performed in a fixed amount of time. The main parameters in GRASP are two: one related to the stopping criterion and another to the quality of the elements in the restricted candidate list.

GRASP may be viewed as a repetitive sampling technique that in each iteration produces a sample solution from an unknown distribution, whose mean and variance are functions of the restrictive nature of the RCL. For example, if RCL is restricted to a single element, then only one solution will be produced and the variance of the distribution will be zero and the mean will equal the value of the greedy solution. If the RCL has more than one element, then many different solutions will be generated, implying a larger variance. Since greediness plays a smaller role in this case, the mean solution value may be worse. However, the value of the best solution found outperforms the mean value and in some cases may be optimal.

The parameter that controls the size of the RCL is α, which is a value between zero and one. In basic GRASP implementations, α is fixed throughout the search. However, Prais and Ribeiro (2000) show that using a fixed α-value may hinder finding solutions of higher quality, which could be found if another value were used. They proposed an extension of the basic GRASP procedure, which they call *Reactive GRASP*, in which the α self-tunes and its value is periodically modified according with the quality of the recently observed solutions.

The way we have adapted the GRASP methodology to construct diverse solutions is as follows. Our goal is to construct a permutation of size m, so in each iteration we select a sector from a RCL to be placed at the next available position. The permutations are constructed from left to right, meaning that we start from position 1 and end in position m. Initially, all sectors are in the unassigned list U. For each sector $i \in U$, we calculate the sum of the entries e_{ij} for all $j \in U$. This measures the attractiveness of each sector. The largest attractiveness value of all the unassigned sectors is multiplied by the α parameter. This value represents a threshold that is used to build the RCL. In particular, the RCL consists of all the sectors in U whose attractiveness measure is at least as large as the threshold value. The procedure randomly selects from the RCL the next sector to be assigned.

After the assignment has been made, the list of unassigned sectors is updated and the measure of attractiveness for the sectors in the updated U list is recalculated. The process is repeated m times and the outcome is a feasible solution to the LOP.

Table 4-1 shows a 14-sector example of the LOP[1]. The main diagonal in Table 4-1 does not contain values because these entries do not contribute to the objective function value in the linear ordering problem. The data values in Table 4-1 are all positive, as required in the LOP. As is typical in LOP instances, we use integer values in Table 4-1.

Table 4-1. Data for a 14-sector LOP

	1	2	3	4	5	6	7	8	9	10	11	12	13	14
1	-	0	9	4	0	6	2	3	9	2	4	5	8	9
2	4	-	5	1	6	8	8	5	0	3	2	6	8	2
3	1	0	-	2	1	4	0	1	4	2	3	4	2	3
4	3	0	7	-	3	6	9	5	9	5	4	3	3	6
5	3	7	9	7	-	0	2	8	6	0	1	0	5	7.
6	2	8	4	4	7	-	2	8	0	6	2	0	0	0
7	9	0	0	6	1	7	-	0	0	8	0	8	0	0
8	4	3	5	0	0	6	0	-	6	0	0	7	4	0
9	0	0	7	7	9	0	0	1	-	4	8	6	2	2
10	0	9	4	0	0	0	4	0	0	-	6	5	2	1
11	9	0	0	6	1	3	1	0	0	8	-	4	3	1
12	4	3	5	0	0	0	3	7	8	0	3	-	1	2
13	0	0	7	7	9	1	0	1	9	2	2	0	-	4
14	0	9	4	0	0	0	0	0	0	1	2	2	6	-

Table 4-2. First 10 solutions in P

p	1	2	3	4	5	6	7	8	9	10	11	12	13	14	C_E
1	8	3	11	2	9	13	1	6	4	5	7	14	10	12	302
2	1	9	14	7	5	10	11	8	13	2	6	4	3	12	322
3	9	13	6	10	1	12	5	4	14	2	8	11	3	7	323
4	7	11	2	4	1	6	13	3	8	5	9	14	10	12	341
5	5	6	12	4	8	10	2	9	1	14	13	3	11	7	331
6	9	4	6	11	14	7	1	13	12	5	8	3	2	10	312
7	12	2	11	6	8	4	13	1	3	5	14	9	7	10	317
8	9	6	4	10	14	12	1	2	13	5	8	11	3	7	322
9	5	4	9	2	11	7	8	3	1	13	14	10	6	12	344
10	4	5	11	7	14	8	3	1	10	9	2	6	13	12	322

---

[1] The example_lop file in the Chapter4 folder contains the data in Table 4-2. Additional instances of the LOP can be found in the instances.zip file of the Chapter7 folder.

When using our GRASP-based Diversification Generation Method with
$\alpha = 0.5$, we can generate a P set with a large number of solutions. If we set
PSize = 100 and use the specified α-value, the first 10 solutions in P are the
ones shown in Table 4-2.

Table 4-2 shows the $p(i)$ values. For example, the third solution in this
table is such that sector 9 is in the first position, followed by sectors 13, 6
and so on. The last column in Table 4-2 shows the objective function value
of each solution. The best solution in Table 4-1 is solution 9 with an
objective function value of 344. The best solution in P before the
application of the Improvement Method is solution number 59 with an
objective function value of 358. We don't show all the solutions generated
with the Diversification Generation Method for obvious reasons; however,
when running the code included in the Chapter4 directory of the
accompanying disc, the reader can use the SSPrint_P and SSPrint_RefSet
functions respectively located in the SSP3.c and SSRefSet3.c files to dump
the P set and the *RefSet* to a specified file.

2.1   Computer Code

The LOP has different interpretations that depend on the context where
the problem is solved. However, for the purpose of our discussion we evoke
the interpretation in Economics and therefore refer to the elements of a
permutation as sectors. The SSGenerate_Sol function (see Function 4-1) in
file SSP3.c is an implementation of the Diversification Generator Method
that generates one solution each time it is called.

The method uses a measure of attractiveness for each sector to construct
solutions. Several alternative measures can be used in connection with this
constructive procedure, as shown in Section 4.2 of Chapter 5. In this tutorial
we calculate the weight of a sector and use it as its measure of attractiveness.
The weight of a sector is simply the sum of the e_{ij} elements in its
corresponding row. The weight of sector i, denoted by w_i, is then defined as
follows, assuming that $e_{ii} = 0$:

$$w_i = \sum_{j=1}^{m} e_{ij} .$$

Since the weight value of a sector i does not depend on any given
permutation, it can be calculated off-line (i.e., before the search begins). The
method starts by initializing the weight array that stores the attractiveness of
each sector. This is done by simply copying the row_sum array onto the
weight array. The row sums are calculated in the SSProblem_Definition
function found in the SSMemory3.c file.

Function 4-1 returns `sol`, a newly generated solution. Because this is a constructive method, at each iteration of the main for-loop the jth element `sol[j]` of a solution is determined. This is done by calculating the maximum weight, `max_weight`, of the unassigned sectors (i.e., those that are not in `sol`). The reduced candidate list is stored in the `candidate` array, which contains all the sectors whose weight is at least as large as `alpha*max_weight`. The procedure randomly selects an index a from the reduced candidate list and assigns `candidate[a]` to the solution (i.e., the assignment `sol[j]=candidate[a]` is made). The weights are updated by subtracting the values corresponding to the recently assigned sector.

Function 4-1. SSGenerate_Sol — File: SSP3.c

```
void SSGenerate_Sol(SS *pb,int sol[])
{
    double alpha=0.5; /* Percentage for RCL */
    int i,j,a,max_weight,n_candi,threshold;
    int *assigned,*candidate,*weight;

    candidate = SSInt_array(pb->nvar);
    assigned  = SSInt_array(pb->nvar);
    weight    = SSInt_array(pb->nvar);

    /* Initialize weights */
    for(i=1;i<=pb->nvar;i++)
      weight[i]=pb->row_sum[i];

    for(j=1;j<=pb->nvar;j++)
    {
      /* Maximum weight for unassigned sectors */
      n_candi=0;
      max_weight=-MAXPOSITIVE;
      for(i=1;i<=pb->nvar;i++)
        if(!assigned[i] && weight[i]>max_weight)
          max_weight=weight[i];

      /* Restricted candidate list */
        threshold= (int)(alpha*(double)max_weight);
        for(i=1;i<=pb->nvar;i++)
        if(weight[i]>=threshold && !assigned[i])
          candidate[++n_candi]=i;

      /* Select one element */
```

```
        a=SSGetrandom(1,n_candi);

    /* Assign the element */
    assigned[candidate[a]]=1;
    sol[j]=candidate[a];

    /* Adapt weights */
    for(i=1;i<=pb->nvar;i++)
        weight[i] -= pb->data[i][candidate[a]];
    }
    free(candidate+1);free(weight+1);free(assigned+1);
}
```

3.   IMPROVEMENT METHOD

The Improvement Method is based on the neighborhood search developed for the LOP in Laguna, Martí and Campos (1999). Insertions are used as the primary mechanism to move from one solution to another. We define INSERT_MOVE($p(j)$, i) to consist of deleting $p(j)$ from its current position j inserting it in position i (i.e., between the sectors currently in positions i-1 and i). This operation results in the ordering p', as follows:

$$p' = \begin{cases} (p(1),\ldots,p(i-1),p(j),p(i),\ldots,p(j-1),p(j+1),\ldots,p(m)) & \text{if } i < j \\ (p(1),\ldots,p(j-1),p(j+1),\ldots,p(i),p(j),p(i+1),\ldots,p(m)) & \text{if } i > j \end{cases}$$

Then the objective function value corresponding to p' can be obtained with the following calculation:

$$C_E(p') = \begin{cases} C_E(p) + \sum_{k=i}^{j-1} \left(e_{p(j)p(k)} - e_{p(k)p(j)} \right) & \text{if } i < j \\ C_E(p) + \sum_{k=j+1}^{i} \left(e_{p(k)p(j)} - e_{p(j)p(k)} \right) & \text{if } i > j \end{cases}$$

The complexity associated with evaluating p' is in both cases $O(m)$. The following example illustrates the insertion mechanism using these equations. Suppose that the matrix corresponding to the permutation $p = (1,2,3,4,5,6,7)$ is the one shown in Figure 4-1, and that we desire to evaluate INSERT_MOVE($p(6)$, 2).

$$E(p) = \begin{pmatrix} 0 & 12 & 5 & 3 & 1 & 8 & 3 \\ 6 & 0 & 3 & 6 & 4 & 4 & 2 \\ 8 & 5 & 0 & 5 & 7 & 0 & 3 \\ -2 & 7 & 2 & 0 & -3 & 6 & 0 \\ 8 & 0 & 3 & -1 & 0 & 4 & 1 \\ 9 & 1 & 6 & 2 & 13 & 0 & 4 \\ 2 & 9 & 4 & -5 & 8 & 1 & 0 \end{pmatrix}$$

Figure 4-1. Matrix associated with $p = (1,2,3,4,5,6,7)$

The elements of the matrix that are inside of the rectangles in Figure 4-1 are those needed to evaluate the objective function corresponding to the permutation $p' = (1, 6, 2, 3, 4, 5, 7)$ that results after the insert move has been performed. Since the objective function value for p is $C_E(p) = 78$, the one for p' is simply:

$$C_E(p') = 78 + (1 - 4) + (6 - 0) + (2 - 6) + (13 - 4) = 78 + 8 = 86$$

Using the matrix in Figure 4-1, it can be verified that the best move involving $p(6)$ is in fact INSERT_MOVE($p(6)$, 3), with a corresponding move value of 11. The move value is the difference between the objective function values after and before the move. In mathematical terms:

Move value $= C_E(p') - C_E(p)$.

Laguna, Martí and Campos (1999) considered two neighborhoods, which they identified as N_1 and N_2:

$N_1 = \{p' : $ INSERT_MOVE($p(j)$, i), for $j = 1, ..., m\text{-}1$ and $i = j+1\}$
$N_2 = \{p' : $ INSERT_MOVE($p(j)$, i), for $j = 1, ..., m$ and $i = 1, 2, ..., j\text{-}1, j+1, ..., m\}$

N_1 consists of permutations that are reached by switching the positions of contiguous sectors $p(j)$ and $p(j+1)$. N_2 consists of all permutations resulting from executing general insertion moves, as defined above. Two search strategies can be defined in conjunction with these neighborhoods. The *best* strategy selects the move with the largest move value among all the moves in the neighborhood. The *first* strategy, on the other hand, scans the list of sectors (in the order given by the current permutation) in search for the first sector $p(f)$ whose movement results in a strictly positive move value (i.e., a

move such that $C_E(p') > C_E(p)$). The move selected by the *first* strategy is
then INSERT_MOVE($p(f)$, i^*), where i^* is the position that maximizes $C_E(p')$.
Note that for N_1, $i^* = f+1$, while for N_2, i^* is chosen from $i = 1, 2, ..., f-1, f+1,$
..., m. Therefore, the combination of the *first* strategy and N_1 results in the
fastest neighborhood search that attempts to find the first improving move
available.

In total, there are four ways of combining the selection strategies with the
neighborhood definitions. These combinations result in four neighborhood
searches labeled *first*(N_1), *best*(N_1), *first*(N_2), and *best*(N_2). Computational
experiments carried out by Laguna, Martí and Campos (1999) showed that
the *first*(N_2) neighborhood search was the most effective, as evident in Table
4-3.

Table 4-3. Experimentation with four Improvement Methods

	first(N_1)	*best*(N_1)	*first*(N_2)	*best*(N_2)
Deviation	25.21%	24.16%	0.15%	0.19%
Optima	0	0	11	11

Table 4-3 reports the average deviation from optimality (Deviation) and
the number of optimal solutions found (Optima). The data used to produce
Table 4-3 consist of 49 instances from the problem library LOLIB[2]. These
instances correspond to real input-output matrices for which the optimal
sector orderings are known. The computational effort associated with these
procedures is equivalent and negligible. The *first*(N_2) is the most effective
with *best*(N_2) a close second, regarding both average deviation from
optimality and number of optimal solutions found. These results indicate
that it is somewhat more effective to search for the best move associated
with a given sector instead of searching for the best move considering all
sectors. A possible explanation for this result is that the "slower" ascent to a
local maximum of *first*(N_2) avoids a premature entrapment in an inferior
local optimum. Based on this finding, Campos, et al. (2001) partitioned N_2
into m neighborhoods:

$$N_2^j = \{p' : \text{INSERT\_MOVE}(p(j), i), i = 1, 2, ..., j-1, j+1, ..., m\}.$$

The N_2^j neighborhood is associated with sector $p(j)$, for $j = 1, ..., m$. For
each sector, there are at most $m-1$ relevant elements in the E matrix (i.e.,
those elements that may contribute to the objective function value). The
elements in the main diagonal are excluded because their sum does not
depend on the ordering of the sectors and are typically set to zero. This

---

[2]   The library is in http://www.iwr.uni-heidelberg.de/groups/comopt/software/LOLIB/

indicates that sectors should not be treated equally by a procedure that selects a sector to explore its neighborhood during a local search. We use w_i, the weight of sector i as defined in Section 2 of this chapter, to differentiate the importance of the sectors in the problem.

The weight values are calculated before the search begins because they are constant and independent of changes in the order of rows or columns. The weight values are used to order the sectors so that the local search explores their corresponding neighborhoods starting from the sector with the largest weight to the sector with the smallest weight.

Using the solutions in Table 4-2 as starting points for the application of the Improvement Method, we obtain the solutions in Table 4-4. The improved solutions in Table 4-4 are all better than the solutions in Table 4-2. In fact all 100 solutions originally in P are improved at least 7% and up to 34% after being subjected to the Improvement Method.

Table 4-4. First 10 solutions in P after the application of the Improvement Method

p	1	2	3	4	5	6	7	8	9	10	11	12	13	14	C_E
1	5	2	11	4	8	7	1	13	9	14	6	10	12	3	383
2	5	2	11	4	7	1	14	6	8	13	9	10	12	3	383
3	2	11	7	1	13	6	5	4	8	9	14	10	12	3	387
4	2	11	7	1	13	6	5	4	8	9	10	14	12	3	387
5	6	5	4	10	2	7	8	1	12	14	13	9	3	11	377
6	5	2	11	4	7	8	1	13	9	6	14	10	12	3	383
7	2	11	7	1	13	6	5	4	10	12	8	9	14	3	385
8	2	7	1	13	9	11	6	5	4	14	10	8	12	3	384
9	5	2	4	7	8	1	13	9	14	11	6	10	12	3	383
10	5	2	4	7	8	1	13	9	14	11	6	10	12	3	383

Because our example problem is fairly small and the Improvement Method is quite effective, some solutions in the original P set converge to the same local optima. For example, the best improved solution with objective function value of 387 (see solution 3 in Table 4-4) appears 11 times in the improved P set. In this case, the Diversification Generation Method is executed additional times until reaching the desired *PSize* improved solutions without duplications.

3.1 Computer Code

The SSImprove_solution function (see Function 4-2) starts with the computation of the sol_inv array. This array is the "inverse" of the solution array sol, in the sense that while the solution array stores the index of the sectors in each position, the inverse array stores the position that each sector occupies. In other words, sol_inv[j]=i if and only if sol[i]=j. The function scans the sectors from the sector with the largest w_j value to the one with the smallest. Instead of introducing a weight array, as done in Function 4-1, we work directly with the pb->row_sum array, which is initialized in the SSProblem_Definition function of the SSMemory3.c file.

The scanning order is established with a call to the SSOrder_i function and stored in the order array. At step i of the procedure, variable j contains the index of the ith element in the order array (j = order[i]), current_pos is the current position in sol (current_pos = sol_inv[j]) and new_pos the new insertion position found with a call to the SSFirst_Insert function, which also returns the move value inc. If the execution of the move improves the current solution (i.e., inc > 0), the SSPerform_Move function is called; otherwise, the move is discarded. If any move has been performed after all the sectors have been considered for insertion, the list of sectors is traversed again starting with the first in the order array. If no improving move is found after completing the execution of the do-loop, then the condition *value > old_value fails and the procedure stops.

Function 4-2. SSImprove_solution — File: SSImprove3.c

```
void SSImprove_solution(SS *pb,int sol[],int *value)
{
    int i,j,inc,old_value,current_pos,new_pos;
    int *order,*sol_inv; /* sol[i]=j <--> sol_inv[j]=i */

    sol_inv=SSInt_array(pb->nvar);
    for(i=1;i<=pb->nvar;i++)
      sol_inv[sol[i]]=i;

    /* Order sectors according to row_sum */
    order = SSOrder_i(pb->row_sum,pb->nvar,1);

    do{
      old_value = *value;
      for(i=1;i<=pb->nvar;i++)
      {
        j=order[i];
        current_pos=sol_inv[j];
        inc = SSFirst_Insert(pb,sol,current_pos,&new_pos);
        if(inc > 0) {
          SSPerform_Move(sol,sol_inv,current_pos,new_pos);
          *value += inc;
        }
      }
```

```
    } while(*value > old_value);

    free(order+1);free(sol_inv+1);
}
```

The SSFirst_Insert function (see Function 4-3) computes the position to insert element sol[i]. It scans the list of sectors (in the order given by sol) in search for the first sector whose movement results in a strictly positive move value. The best_inc variable stores the move value when sol[i] is inserted in position. The function starts by initializing position to i-1, except if i is equal to 1. In this case it is initialized to 2. Then, insertions to positions j such that j < i are examined first, followed by those positions j such that j > i. The best position found (the one associated with the largest move value) is stored in best_inc. When an insertion that results in an improved solution is found (best_inc > 0), the function terminates and best_inc is returned.

Function 4-3. SSFirst_Insert — File: SSImprove3.c

```
int SSFirst_Insert(SS *pb, int sol[], int i, int *position)
{
    int k,best_inc,inc;

    /* Initialize best_inc */
    if(i>2) {
        best_inc  = pb->data_dif[sol[i]][sol[i-1]];
        *position = i-1;
    }
    else {
        best_inc  = pb->data_dif[sol[i+1]][sol[i]];
        *position = i+1;
    }
    if(best_inc>0) return best_inc;

    /* Insert sol[i] in a position less than i */
    inc=0;
    for(k=i-1;k>=1;k--)
    {
        inc += pb->data_dif[sol[i]][sol[k]];
        if(inc > best_inc)
        {
          best_inc = inc;
            *position = k;
```

```
        if(best_inc>0)
            return best_inc;
    }
}

/* Insert sol[i] in a position greater than i */
inc=0;
for(k=i+1;k<=pb->nvar;k++)
{
    inc += pb->data_dif[sol[k]][sol[i]];
    if(inc > best_inc)
    {
    best_inc  = inc;
        *position = k;
    if(best_inc>0) return best_inc;
    }
}
return best_inc;
}
```

4.   REFERENCE SET UPDATE METHOD

The Reference Set Update Method is essentially the same as the one described in Chapters 2 and 3. The size of the reference set in this tutorial is set to 10 (i.e., $b = 10$), where initially $b_1 = 5$ solutions are selected due to their high quality and $b_2 = 5$ solutions are selected due to their diversity. The result of constructing the reference with the improved solutions in P is shown in Table 4-5.

Table 4-5. Initial reference set

p	1	2	3	4	5	6	7	8	9	10	11	12	13	14	C_E
7	2	11	7	1	13	6	5	4	8	9	10	12	14	3	387
18	2	11	7	1	13	6	5	4	8	9	14	10	12	3	387
32	2	11	7	1	13	6	5	4	8	9	10	14	12	3	387
1	2	7	1	13	6	5	4	8	9	14	3	11	10	12	386
3	2	7	1	13	6	5	4	8	9	10	14	3	11	12	386
24	2	7	1	13	9	11	6	5	4	10	12	14	8	3	384
30	5	2	4	7	8	1	14	13	9	3	11	6	10	12	383
10	5	2	11	4	7	1	6	10	12	14	8	13	9	3	382
5	7	1	6	5	4	10	14	2	12	8	13	9	3	11	380
40	5	1	4	8	9	14	2	11	13	7	6	10	12	3	373

The first column in Table 4-5 shows the solution number corresponding to the P set that contains 100 improved solutions without duplications. The first three solutions (labeled 7, 18 and 32), that were added to the reference set because of their quality, are almost identical. They differ in the order of sectors 10, 12 and 14, which occupy positions 11, 12 and 13 in the permutation. So, strictly speaking, these three solutions are different from each other. The Combination Method in the following section, however, would not produce new trial solutions when combining the first three solutions in the reference set among themselves. In Chapter 5 we explore a strategy that avoids this premature convergence of the reference set by establishing a minimum diversity for the b_1 solutions that are chosen due to their quality.

5. COMBINATION METHOD

As noted before, the Combination Method is an element of scatter search whose design depends on the problem context. Although it is possible in some cases to design context-independent combination procedures (as shown in Chapter 7), it is generally more effective to base the design on specific characteristics of the problem setting. The Combination Method in this tutorial employs a min-max construction based on votes.

The method scans (from left to right) each reference permutation in a subset, and uses the rule that each reference permutation in the combination subset votes for its first sector that so far has not been included in the combined permutation (referred to as the "incipient sector"). The voting determines the sector to be assigned to the next free position in the combined permutation. This is a min-max rule in the sense that if any sector of the reference permutation is chosen other than the incipient sector, then it would increase the deviation between the reference and the combined permutations. Similarly, if the incipient sector were placed later in the combined permutation than its next available position, this deviation would also increase. So the rule attempts to minimize the maximum deviation of the combined solution from the reference solution, subject to the fact that other reference solutions in the subset are also competing to contribute.

This voting scheme can be implemented using a couple of variations that depend on the way votes are modified:

1. The vote of a given reference solution is weighted according to the incipient sector's position (referred to as the "incipient position"). A smaller incipient position gets a higher vote. For example, if the sector in the first position of some reference permutation is not assigned to the combined permutation during the first 4 assignments, then the vote is weighted more heavily to increase the chances of having that sector

assigned to position 5 of the combined permutation. The rule emphasizes the preference of this assignment to one that introduces an (incipient) sector that occurs later in some other reference permutation.

2. A bias factor gives more weight to the vote of a reference permutation with higher quality. Within the current organization of the scatter search implementation in this tutorial such a factor should be very slight because it is expected that high quality solutions will be strongly represented anyway.

We chose to implement the first variant with a tie-breaking rule based on solution quality. The tie-breaking rule is used when more than one sector receives the same votes. Then the sector with highest weighted vote is selected, where the weight of a vote is directly proportional to the objective function value of the corresponding reference solution.

An example of combining solutions 18 and 5 in Table 4-5 is shown in Table 4-6. The first column shows the step number, indicating that a new trial solution is constructed in $m = 14$ steps. Each reference solution (18 and 5) has two columns: the sector index and the vote value.

Table 4-6. Combination of solutions 18 and 5 of the reference set

| | Solution 18 | | Solution 5 | | | | New |
Step	Sector	Vote	Sector	Vote	Winner	Criterion	solution
1	2	1	7	1	Sol. 18	obj	2
2	11	1	7	2	Sol. 5	vote	7
3	11	2	1	2	Sol. 18	obj	11
4	1	1	1	3	Sol. 5	vote	1
5	13	1	6	3	Sol. 5	vote	6
6	13	2	5	3	Sol. 5	vote	5
7	13	3	4	3	Sol. 18	obj	13
8	4	1	4	4	Sol. 5	vote	4
9	8	1	10	4	Sol. 5	vote	10
10	8	2	14	4	Sol. 5	vote	14
11	8	3	12	3	Sol. 18	obj	8
12	9	3	12	4	Sol. 5	vote	12
13	9	4	9	2	Sol. 18	vote	9
14	3	1	3	2	Sol. 5	vote	3

In step 1 of Table 4-6, solution 18 votes for sector 2 to go in the first position of the new solution. Solution 5 votes for sector 7. Hence, sector 2 and 7 receive one vote and all other sectors receive zero votes, the tie-breaking rule is used and sector 2 is chosen because solution 18, which is the second in the reference set, has a better objective function value than solution 5, which is the ninth in the reference set. The "Winner" column in Table 4-6 shows the solution that provides the next sector to be included in the new solution and the "Criterion" column shows whether the selection was made due to the number of votes or to the tie-breaking rule that uses the objective function value of the reference solutions.

5.1 Computer Code

The SSCombine function in the RefSet3.c file implements the Combination Method based on votes for the special case of two reference solutions. The function has six arguments: the pointer pb to the SS structure, the index of the reference solutions sol1 and sol2, and their corresponding objective function values value1 and value2. The last argument newsol is used to return the generated solution.

The i1 variable stores the index of the first sector in sol1 not currently included in newsol. That is, i1 is the incipient sector of sol1. Similarly, index i2 is the incipient sector of solution sol2. At each iteration i of the for-loop, the element newsol[i] is determined. First, the votes v1 of sol1 and v2 of sol2 are computed and then the solution with the larger vote value is selected and its incipient sector is assigned to newsol[i]. If both reference solutions have the same vote value, the one with the larger objective function value is selected. The details about how to compute the votes can be found in Function 4-4. The assign array flags the sectors currently in newsol. The array is used to update the incumbent sector at each iteration.

Function 4-4. SSCombine — File: SSRefSet3.c

```
void SSCombine(SS *pb, int sol1[], int sol2[], int value1,
               int value2, int newsol[])
{
    int i,a;
    int i1=1;      /* Index in sol1 of its incipient sector */
    int i2=1;      /* Index in sol2 of its incipient sector */
    int v1,v2;     /* Votes for sol1[i1] and sol2[i2]       */
    int *assign;   /* Assigned sectors                      */

    assign = SSInt_array(pb->nvar);
    for(i=1;i<=pb->nvar;i++)
    {
        if(i1>pb->nvar) a=sol2[i2];
        else if (i2>pb->nvar) a=sol1[i1];
        else
        {
            v1 = 1 + i - i1;
```

```
            v2 = 1 + i - i2;

            if(v1>v2)             a=sol1[i1];
            else if(v2>v1)        a=sol2[i2];
            else if(value1>value2) a=sol1[i1];
            else                  a=sol2[i2];
        }
        assign[a]=1;
        newsol[i]=a;

        while(i1<=pb->nvar && assign[sol1[i1]]) i1++;
        while(i2<=pb->nvar && assign[sol2[i2]]) i2++;
    }
    free(assign+1);
}
```

6.   SUMMARY OF C FUNCTIONS

Functions in file **SSMemory.c**

```
double *SSDouble_array(int size);
double **SSDouble_matrix(int nrows,int ncolumns);
void   SSFree_DataStructures(SS *pb);
void   SSFree_double_matrix(double **matrix,int nrows);
void   SSFree_int_matrix(int **matrix,int nrows);
int    *SSInt_array(int size);
int    **SSInt_matrix(int nrows,int ncolumns);
SS     *SSProblem_Definition(char *filename,int b,
                             int PSize);
int    **SSReadData(char *file,SS *pb);
```

Functions in file **SSRefSet.c**

```
void   SSCombine(SS *pb,int sol1[],int sol2[],int value1,
                 int value2,int newsol[]);
void   SSCombine_RefSet(SS *pb);
void   SSCreate_RefSet(SS *pb);
void   SSPrint_RefSet(SS *pb);
void   SSRebuild_RefSet(SS *pb);
void   SSTryAdd_RefSet(SS *pb,int sol[],int value);
void   SSUpdate_RefSet(SS *pb);
```

Functions in file **SSP.c**

```
void    SSCreate_P(SS *pb);
void    SSGenerate_Sol(SS *pb,int sol[]);
void    SSPrint_P(SS *pb);
```

Functions in file **SSImprove3.c**

```
int     SSFirst_Insert(SS *pb,int sol[],int i,int *position);
void    SSImprove_solution(SS *pb,int sol[],int *value);
void    SSPerform_Move(int sol[],int invsol[],int i,
                       int position);
```

Functions in file **SSTools.c**

```
void    SSAbort(char texto[]);
void    SSBestSol(SS *pb,int sol[],int *value);
int     SSDist_RefSet(SS *pb,int num,int sol[]);
int     SSEqualSol(int sol1[],int sol2[],int dim);
int     SSIsInRefSet(SS *pb,int sol[]);
int     SSMax_dist_index(SS *pb,int dist[]);
int     *SSOrder_d(double weights[],int num,int type);
int     *SSOrder_i(int weights[],int num,int type);
float   SSRandNum(void);
void    SSUpdate_distances(SS *pb,int min_dist[],int rs_index);
```

Chapter 5

ADVANCED SCATTER SEARCH DESIGNS

Because life is too short to drink cheap beer

Warsteiner, made since 1753 according to German Purity Law

The preceding three tutorial chapters described somewhat basic implementations of scatter search. The purpose was to introduce basic scatter search strategies in three different problem settings. However, not all the elements implemented in the tutorial chapter can be considered basic. For example, the use of frequency information in the Diversification Generation Method of Chapter 2 introduces a memory-based procedure that advanced designs described in Chapter 6 exploit. Another instance of a strategy that goes beyond a basic implementation is the use of GRASP constructions to generate diverse solutions in Chapter 4. Likewise, the Improvement Method in Chapter 3, which is capable of dealing with feasible and unfeasible initial solutions, is typical of advanced scatter search implementations.

In some areas, however, the tutorial chapters employed basic designs. Advanced strategies for the manipulation of the reference set were purposefully avoided in the tutorial chapters. Similarly, the Combination Method was designed to operate on two reference solutions only. While these basic schemes might actually "do the job" in a number of settings, some situations warrant the addition of advanced mechanisms with the goal of improving performance (as measure by the expected quality of solutions found in a desired computational effort).

When considering advanced strategies in a metaheuristic framework, the goal of improving performance often conflicts with the goal of designing a procedure that is easy to implement and fine tune. Advanced designs generally translate into higher complexity and additional search parameters. This is certainly true for some of the designs that we discuss in this chapter, but not for all. We have organized the chapter around advanced strategies associated with each scatter search method without suggesting a particular order in which they should be added to basic implementations. The structure of the chapter simply reflects our belief that there is no simple recipe that can be used to follow a predetermined order in which advanced strategies should be added to progressively improve the performance of a scatter search implementation.

Because metaheuristic procedures are widely applied to combinatorial optimization problems, we focus on the linear ordering problem (LOP) as a means to illustrate the implementation of the advanced strategies described in this chapter. Using the LOP for illustration purposes does not limit our ability to generalize, because important elements, such as the Reference Set Update Method, does not depend on the application context. Table 5-1 summarizes the context-dependent elements in the scatter search implementations in Chapters 2, 3 and 4.

Table 5-1. Context-dependent elements

Element	Nonlinear Optimization	Knapsack Problems	Linear Ordering Problems
Diversification Generation Method	Frequency counts	Systematic	GRASP constructions
Improvement Method	Simplex	Greedy switching	Insertions
Combination Method	Linear combinations	Scores	Voting
Distance Measure	Euclidean	Absolute difference	Absolute difference
Solution Representation	Continuous variables	Binary variables	Permutation vectors

In Table 5-1 we can observe that the Diversification Generation, Improvement and Combination Methods are context dependent. Generating diversification effectively depends on the solution representation, which in turn depends on the problem being solved. Distance measures, an essential element of the Diversification Generation Method, also depend on the solution representation.

1.   REFERENCE SET

The reference set is the heart of a scatter search procedure. If at any given time during the search all the reference solutions are alike, as measured by an appropriate metric, the scatter search procedure will most likely be incapable of improving upon the best solution found even when employing a sophisticated procedure to perform combinations or improve new trial solutions. The Combination Method is limited by the reference solutions that it uses as input. Hence, having the most advanced Combination Method is of little advantage if the reference set is not carefully built and maintain during the search.

In this section, we explore advanced mechanisms associated with the reference set. Particularly, we look at dynamic updating of the reference set, the use of multi-tier reference set and the control of diversity.

1.1   Dynamic Updating

In the basic design, the new solutions that become members of *RefSet* are not combined until all pairs in *NewSubsets* are subjected to the Combination Method (see Figure 1-4 in Chapter 1). In the tutorial chapters, we introduced the notion of *Pool* of trial solutions (see Section 3 in Chapter 2). The new reference set is built with the best solutions in the union of *Pool* and the solutions currently in *RefSet*. This strategy is called the *Static Update* of the reference set.

The alternative to the static update is the *Dynamic Update* strategy, which applies the Combination Method to new solutions in a manner that combines new solutions faster than in the basic design. That is, if a new solution is admitted to the reference set, the goal is to allow this new solution to be subjected to the Combination Method as quickly as possible. In other words, instead of waiting until all the combinations have been performed to update the reference set, if a new trial solution warrants admission in the reference set, the set is immediately updated before the next combination is performed.

Figure 5-1 illustrates this mechanism with a hypothetical reference set containing 4 solutions. This figure shows that the dynamic updating of the reference set may result in the replacement of solutions that have not been combined. In particular, Figure 5-1 depicts a *RefSet* consisting of solutions x^1, x^2, x^3, and x^4, which have been ordered from best to worst according to their objective function value $f(x)$. The figure also depicts the combination of the pair (x^1, x^2), which after the application of the Improvement Method results in the solution labeled y. This new trial solutions is such that $f(x^2) < f(y) < f(x^3)$ in a hypothetical minimization problem.

After the first combination the reference set is changed in such a way that the updated *RefSet* consists of solutions x^1, x^2, y, and x^3. The search continues by combining the pair (x^1, y) instead of the combination of the (x^1, x^3) pair that would have been made under the static updating. The (x^1, y) combination might cause additional changes to *RefSet*, but even if it doesn't the combination of the pair (x^1, x^4) has been eliminated from consideration.

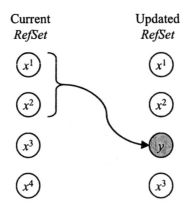

Current   Updated
RefSet   *RefSet*

Figure 5-1. Dynamic Reference Set Update

The advantage of the dynamic update is that if the reference set contains solutions of inferior quality, these solutions are quickly replaced and future combinations are made with improved solutions. The disadvantage is that some potentially promising combinations are eliminated before they can be considered. The implementation of dynamic updating is more complex than its static counterpart. Also, in the static update the order in which the combinations are performed is not important because the *RefSet* is not updated until all combinations have been performed. In the dynamic updating, the order is quite important because it determines the elimination of some potential combinations. Hence, when implementing a dynamic update of the reference set, it may be necessary to experiment with different combination orders as part of the fine tuning of the procedure.

The implementation of the dynamic update strategy in the context of the linear ordering problem is shown in Function 5-1.

Function 5-1. SSDynamicUpdate_RefSet — File: SSRefSet3a.c

```
void SSDynamicUpdate_RefSet(SS *pb)
{
    int i,j,a,b,value;
    int *newsol;

    newsol = SSInt_array(pb->nvar);
    pb->rs->NewSolutions=0;

    for(i=1;i<pb->rs->b;i++)
    for(j=i+1;j<=pb->rs->b;j++)
    {
        a=pb->rs->order[i];
```

```
    b=pb->rs->order[j];
    if( (a<b && pb->rs->comb[a][b]) ||
        (a>b && pb->rs->comb[b][a]) )
    {
      if(a<b) pb->rs->comb[a][b]=0;
      else     pb->rs->comb[b][a]=0;
      SSCombine(pb,pb->rs->sol[a],pb->rs->sol[b],
               pb->rs->ObjVal[a],pb->rs->ObjVal[b],newsol);
      value=sol_value(pb,newsol);
      SSImprove_solution(pb,newsol,&value);
      SSTryAdd_RefSet(pb,newsol,value);
    }
  }
  free(newsol+1);
  pb->CurrentIter++;
}
```

The `comb` matrix in Function 5-1 has been added to our original REFSET structure to control whether two solutions have been already combined or not. The element `comb[a][b]` is meaningful only if a < b. If `comb[a][b]` = 1 the procedure allows the combination of the reference solutions `pb->rs->sol[a]` and `pb->rs->sol[b]`. A newly generated solution is stored in `newsol`. New trial solutions are subjected to the Improvement Method and then immediately submitted to the `SSTryAdd_RefSet` function, which checks whether the improved trial solution qualifies for admission in the *RefSet*. Therefore, in this dynamic implementation there is no intermediate *Pool* of solutions as in the static design. Recall that *Pool* stores all the improved new trial solutions generated with the application of the Combination and Improvement Methods before they are submitted to the `SSTryAdd_RefSet` function. The complete code of this implementation can be found in the Chapter5\Section 1.1 folder of the accompanying compact disc.

1.2   Rebuilding and Multi-Tier Update

In basic scatter search implementations, the reference set is updated by replacing the reference solution with the worst objective function value with a new trial solution that has a better objective function value. Since we consider that *RefSet* is always ordered, the best solution is x^1 and the worst solution is x^b. So, when a new trial solution x is generated as a result of the application of the Combination and Improvement Methods, the objective function value of the new trial solution is used to determine whether the *RefSet* needs to be updated. In a minimization problem, if:

$$x \notin RefSet \text{ and } f(x) < f(x^b),$$

RefSet is updated by making $x^b = x$ and reordering the set. This updating procedure treats all the solutions in the reference set in the same way by measuring their merit with their corresponding objective function values. We now explore mechanisms that differentiate solutions using additional measure of merit that are not based on the objective function value.

1.2.1   Rebuilding

In Chapter 2, we introduced an updating procedure that is triggered when no new trail solutions are admitted to the reference set. This update adds a mechanism to partially rebuild the reference set when the Combination and Improvement Methods do not provide solutions of sufficient quality to displace current reference solutions.

The *RefSet* is partially rebuilt with a diversification update that works as follows and assumes that the size of the reference set is $b = b_1 + b_2$. Solutions $x^{b_1+1}, \ldots, x^b$ are deleted from *RefSet*. The Diversification Generation Method is reinitialized considering that the goal is to generate solutions that are diverse with respect to the reference solutions $x^1, \ldots, x^{b_1}$. Then, the Diversification Generation Method is used to construct a set P of new solutions. The b_2 solutions $x^{b_1+1}, \ldots, x^b$ in *RefSet* are sequentially selected from P with the criterion of maximizing the minimum distance, where the distance measure is defined in the context of the problem being solved. The min-max criterion, which is part of the Reference Set Update Method, is applied with respect to solutions $x^1, \ldots, x^{b_1}$ when selecting solution x^{b_1+1}, then it is applied with respect to solutions $x^1, \ldots, x^{b_1+1}$ when selecting solution x^{b_1+2}, and so on. A schematic representation of the rebuilding operation is depicted in Figure 5-2.

The computer code for this updating procedure is Function 2-6 in Chapter 2 for unconstrained nonlinear optimization. Similar functions for the knapsack problem and the linear ordering problem can be respectively found in the Chapter3 and Chapter4 folders in the compact disc included in this book.

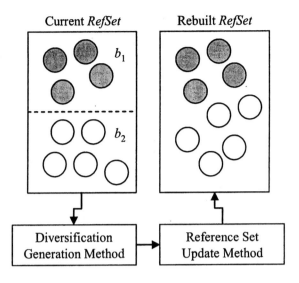

Figure 5-2. Rebuilding strategy

1.2.2   2-Tier Update

Instead of waiting until the reference set has converged, that is, it has reached a state in which no new solutions are admitted, an updating procedure that proactively injects diversification into the search can be used. The updating procedure employs a 2-tier design, where the first tier *RefSet₁* consists of b_1 high quality solutions and *RefSet₂* consists of b_2 diverse solutions. The update has the goal of dynamically preserving diversity in the reference set, instead of allowing it to become homogenous by only admitting high quality solutions that in some applications tend to be very similar to each other. Hence, in addition to updating the reference set when new trial solutions of high quality are found with the Combination and Improvement Methods, the reference set is also updated with highly diverse solutions.

Specifically, the update consists of partitioning the reference into two subsets:

$$RefSet_1 = \left\{ x^1, \dots, x^{b_1} \right\}$$

$$RefSet_2 = \left\{ x^{b_1+1}, \dots, x^b \right\}$$

The first subset is referred to as the "quality" subset and the second is referred to as the "diverse" subset. The solutions in *RefSet₁* are ordered according to their objective function value and the set is updated with the goal of increasing quality, using the criterion of the basic scatter search design. That is, a new solution x replaces reference solution x^{b_1} if $f(x) < f(x^{b_1})$ in a minimization problem. The solutions in *RefSet₂* are ordered according to their diversity value and the update has the goal of increasing diversity. Therefore, a new solution x replaces reference solution x^b if $d_{\min}(x) > d_{\min}(x^b)$. Figure 5-3 shows a schematic representation of the 2-tier reference set update.

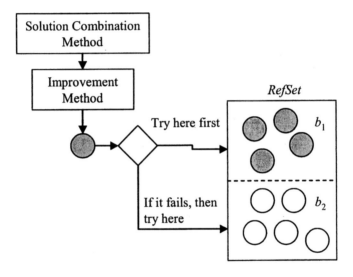

Figure 5-3. Schematic representation of the 2-tier update

The implementation of the 2-tier update strategy in the context of the linear ordering problem is given in folder Chapter5\Section 1.2\2Tier. Function 5-2 shows the code that implements a Reference Set Update Method for a 2-tier *RefSet*.

Function 5-2. SSTryAdd_RefSets — File: SSRefSet3b.c

```
void SSTryAdd_RefSets(SS *pb, int sol[], int value)
{
    int i,j,k,worst_index1,worst_index2,dist;

    worst_index1 = pb->rs1->order[pb->rs1->b];
    worst_index2 = pb->rs2->order[pb->rs2->b];
    dist = SSDist_RefSet(pb,pb->rs1->b,sol);
```

```
/* Try to add to RefSet1 according to the Obj Value */
if(!SSIsInRefSet(pb,pb->rs1,sol) &&
    value > pb->rs1->ObjVal[worst_index1])
{
  /* Find position i for solution insertion */
  i=pb->rs1->b;
  while(i>=1 && value>pb->rs1->ObjVal[pb->rs1->order[i]])
    i--;
  i++;

  /* Replace solution */
  for(j=1;j<=pb->nvar;j++)
    pb->rs1->sol[worst_index1][j]=sol[j];
  pb->rs1->ObjVal[worst_index1]=value;
  pb->rs1->iter[worst_index1]=pb->CurrentIter;

  /* Reorder RefSet1 */
  for(j=pb->rs1->b;j>i;j--)
    pb->rs1->order[j]=pb->rs1->order[j-1];
  pb->rs1->order[i]=worst_index1;

  /* Update distances in RefSet2 */
  for(k=1;k<=pb->rs2->b;k++)
  {
    dist=0;
    for(j=1;j<=pb->nvar;j++)
      dist += abs(sol[j]-pb->rs2->sol[k][j]);
    if(pb->rs2->DivVal[k]> dist)
      pb->rs2->DivVal[k]=dist;
  }
  pb->rs1->NewSolutions=1;
}

/* Try to add to RefSet2 according to Diversity Value */
else if(!SSIsInRefSet(pb,pb->rs2,sol) &&
        dist>pb->rs2->DivVal[worst_index2])
{
  /* Find position i for solution insertion */
  i=pb->rs2->b;
  while(i>=1 && dist>pb->rs2->DivVal[pb->rs2->order[i]])
    i--;
  i++;
```

```
/* Replace solution */
for(j=1;j<=pb->nvar;j++)
  pb->rs2->sol[worst_index2][j]=sol[j];
pb->rs2->ObjVal[worst_index2]=value;
pb->rs2->DivVal[worst_index2]=dist;
pb->rs2->iter[worst_index2]=pb->CurrentIter;

/* Reorder ReSet */
for(j=pb->rs2->b;j>i;j--)
  pb->rs2->order[j]=pb->rs2->order[j-1];
pb->rs2->order[i]=worst_index2;

pb->rs2->NewSolutions=1;
}
}
```

In Function 5-2, rs1 points to *RefSet₁* and rs2 points to *RefSet₂*. A new trial solution (newsol) is submitted to the SSTryAdd_RefSets function, which first tests for admission to *RefSet₁* (first if statement) and upon failure then tests for admission to *RefSet₂* (if else statement). The first part of the function (the one within the first if statement) is similar to the SSTryAdd_RefSet function used in the tutorial chapters, with the difference that when a solution is added to *RefSet₁* the distance values of the solutions in *RefSet₂* (pb->rs2->DivVal) must be updated. We have included the DivVal array in the REFSET structure to store the d_{min} value of solutions in *RefSet₂*. The second part of the function (the one within the if-else statement) performs the update of *RefSet₂* when a solution is added to this set. This code employs a slightly different implementation of the SSIsInRefSet function, in which the pointer of *RefSet* is also passed as an argument. To simplify the implementation, we have introduced the worst_index1 variable to store the index of the worst solution in *RefSet₁* according to its objective function value and the worst_index2 variable to store the index of the worst solution in *RefSet₂* according to its distance value.

The 2-tier update can be used in combination with the rebuilding mechanism. The implementation is straightforward by keeping *RefSet₁* and reinitializing the Diversification Generation Method in order to rebuild *RefSet₂* with solutions that are diverse among them and with respect to *RefSet₁*. The SSRebuild_RefSet2 function implements this updating mechanism. This function can be found in the SSRefSet3b.c file of the Chapter5\Section 1.2\2Tier directory.

The Combination Method based on votes introduced in Section 5 of Chapter 4 uses the objective function value as a tie-breaking criterion to determine the origin of the next element to be included in the new trial solution (see Function 4-4). An alternative design in the context of a 2-tier *RefSet* is to use the diversity value d_{min} when the reference solutions being combined are members of *RefSet₂*. Another variant would be to use the original trial solution instead of its corresponding improved solution when adding solutions to *RefSet₂*. This strategy would tend to increase the diversity of *RefSet₂*.

1.2.3   3-Tier Update

This update is an extension of the 2-tier update that maintains a list of the "best generators". A "good generator" is a reference solution that generates high quality trial solutions when used as input to the Combination Method. The 3-tier update uses a reference set of size $b = b_1 + b_2 + b_3$, which is divided into the following three subsets:

$$RefSet_1 = \left\{ x^1, \ldots, x^{b_1} \right\}$$

$$RefSet_2 = \left\{ x^{b_1+1}, \ldots, x^{b_1+b_2} \right\}$$

$$RefSet_3 = \left\{ x^{b_1+b_2+1}, \ldots, x^b \right\}$$

RefSet₁ and *RefSet₂* are updated using the same rules as in the 2-tier update. In order to update *RefSet₃*, we keep track of $g(x)$, the objective function value of the best solution ever created from a combination of $x \in RefSet_1$ and any other reference solution. *RefSet₃* is ordered according to $g(x)$ in such a way that $g\left(x^{b_1+b_2+1}\right) < g\left(x^{b_1+b_2+2}\right) < \cdots < g\left(x^b\right)$ for a minimization problem. When x^{b_i} in *RefSet₁* is replaced with a newly created solution of higher quality, we compare $g\left(x^{b_i}\right)$ with $g\left(x^b\right)$ and update *RefSet₃* if appropriate.

This design would be particularly helpful in settings where solutions of relatively low quality are capable of producing high quality solutions. The design would allow this "good generators" to participate in additional combinations once they have been replaced from *RefSet₁*. The initialization of *RefSet₁* and *RefSet₂* is the same as in the 2-tier design. *RefSet₃* is initialized with the best solutions in *P* that were not included in *RefSet₁*.

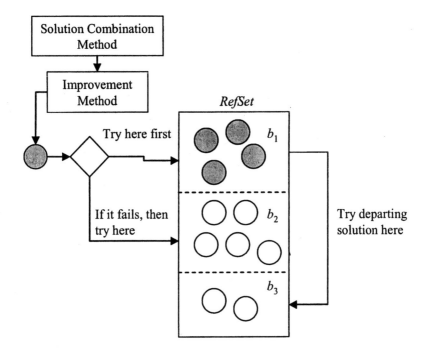

Figure 5-4. Schematic representation of the 3-tier update

Figure 5-4 shows that the update of *RefSet₃* depends on the update of *RefSet₁*, in the sense that *RefSet₃* can only be updated if a solution in *RefSet₁* is replaced. If no solutions are admitted to *RefSet₁* or *RefSet₂* the search ends or the rebuilding procedure of Section 1.2.1 of this chapter may be applied. This would mean that *RefSet₁* and *RefSet₃* are kept and *RefSet₂* is rebuilt with the Diversification Generation Method.

The implementation of the 3-tier update strategy in the context of the linear ordering problem is shown in Function 5-3.

Function 5-3. SSTryAdd_RefSets — File: SSRefSet3c.c

```
void SSTryAdd_RefSets(SS *pb,int sol[],int value)
{
    int i,j,k,dist,g_value,changedist,*rs_order;
    int worst_index1,worst_index2,worst_index3;

    worst_index1 = pb->rs1->order[pb->rs1->b];
    worst_index2 = pb->rs2->order[pb->rs2->b];
    worst_index3 = pb->rs3->order[pb->rs3->b];
    dist    = SSDist_RefSet(pb,pb->rs1->b,sol);

    /* Try to add to RefSet1 according to Obj. Value */
```

```
if(!SSIsInRefSet(pb,pb->rs1,sol) &&
   value>pb->rs1->ObjVal[worst_index1])
{
  /* Try to add the sol. removed from RefSet1 to RefSet3 */
  if(pb->rs1->g[worst_index1] > pb->rs3->g[worst_index3])
  {
    /* Find position i for sol. insertion */
    g_value=pb->rs1->g[worst_index1];
    i=pb->rs3->b;
    while(i>=1 && g_value>pb->rs3->g[pb->rs3->order[i]])
      i--;
    i++;
    SSInsertInRefSet(pb,pb->rs3,sol,worst_index3,i,value,0);
    pb->rs3->g[worst_index3]=g_value;
  }

  /* Find position i in RefSet1 for sol. insertion */
  i=pb->rs1->b;
  while(i>=1 && value>pb->rs1->ObjVal[pb->rs1->order[i]])
    i--;
  i++;
  SSInsertInRefSet(pb,pb->rs1,sol,worst_index1,i,value,0);

  /*Update distances and reorder RefSet2 */
  changedist=0;
  for(k=1;k<=pb->rs2->b;k++)
  {
      dist=0;
      for(j=1;j<=pb->nvar;j++)
        dist += abs(sol[j]-pb->rs2->sol[k][j]);
      if(pb->rs2->DivVal[k]> dist) {
        changedist=1;
        pb->rs2->DivVal[k]=dist;
      }
  }
  if(changedist)
  {
      rs_order = SSOrder_i(pb->rs2->DivVal,pb->rs2->b,1);
      for(k=1;k<=pb->rs3->b;k++)
        pb->rs2->order[k]=rs_order[k];
      free(rs_order+1);
  }
}
```

```
/* Try to add to RefSet2 according to Divers. Value */
else if(!SSIsInRefSet(pb,pb->rs2,sol) &&
        dist>pb->rs2->DivVal[worst_index2])
{
   /* Find position i for sol. insertion */
   i=pb->rs2->b;
   while(i>=1 && dist>pb->rs2->DivVal[pb->rs2->order[i]])
      i--;
   i++;
   SSInsertInRefSet(pb,pb->rs2,sol,worst_index2,i,value,dist);
}
}
```

Following the notation introduced in the previous section, rs1 points to *RefSet₁* and rs2 points to *RefSet₂*. We have added the g array in the REFSET structure to store the $g(x)$ value of solutions in *RefSet₁* and *RefSet₃*. We have also added the worst_index3 variable to store the index of the worst solution in *RefSet₃* according to its g-value. The insertion of solutions to a reference set and the reordering are performed with a call to the SSInsertInRefSet function.

Function 5-3 operates in a similar way as Function 5-2 with the difference that when a solution is going to be added to *RefSet₁* in the first if statement and before deleting the worst solution in this set (pb->rs1->sol[worst_index1]) the procedure attempts to add it to *RefSet₃*. The expression pb->rs1->g[worst_index1] > pb->rs3->g[worst_index3] compares the g-value of the solution leaving *RefSet₁* with the g-value of the worst solution in *RefSet₃*.

We do not discuss the code for the SSCombine_RefSets function in file SSRefSet3c.c because it has 120 lines. We point out, however, that in the 3-tier *RefSet* context this function implements the combinations of all pairs of elements within each reference set and between the three different reference sets. All the files related to the implementation of the 3-tier reference set strategy can be found in the Chapter5\Section 1.2\3Tier folder.

1.3   Solution Duplication and Diversity Control

In Section 4 of Chapter 4, we observed that the high quality solutions that are used to build the initial reference set may be very similar to each other. For instance, the first three solutions in Table 4-5 have the same objective function value and they differ from each other only in the position of three out of fourteen elements in the permutation. When these solutions are used as input to the Combination Method, they do not posses enough diversity to generate new trial solutions. The strategies in this section are designed to control the amount of diversity among the high quality solutions admitted to the reference set.

1.3.1 Minimum Diversity Test

Basic scatter search implementations are designed to check that the reference set does not contain duplications; however they generally do not check how diverse the b_1 quality solutions are when creating the initial *RefSet*. On the other hand, recall that the b_2 diverse solutions are subjected to a strict diversity check with the min-max criterion. Hence, the minimum diversity test refers only to the b_1 high quality solutions chosen as members of the initial *RefSet*.

The minimum diversity test operates as follows. After the P set has been created, the best solution according to the objective function value is selected to become x^1 in the reference set. Then, x^1 is deleted from P and the next best solution x in P is chosen and added to *RefSet* only if

$$d_{min}(x) \geq th\_dist.$$

In other words, at each step we add the next best solution in P only if the minimum distance between the chosen solution x and the solutions currently in *RefSet* is at least as large as the threshold value *th_dist*. This criterion is used during the first b_1 solution selections. Function 5-4 implements the minimum diversity test.

Function 5-4. SSCreate_RefSet — File: SSRefSet3d.c

```
void SSCreate_RefSet(SS *pb)
{
    int b1,i,j,a,num,th_dist=10;
    int *p_order,*rs_order,*min_dist;

    /* Order solutions in P */
    p_order = SSOrder_i(pb->p->ObjVal,pb->p->PSize,1);

    /* Add the best b1 solutions in P to RefSet */
    b1 = pb->rs->b / 2;
    i=num=0;
    while(num<b1)
    {
        i++;
```

```
  if(i>pb->p->PSize)
    SSAbort("Reduce the Diversity Test Value");

  /* Minimum diversity test */
  if(SSDist_RefSet(pb,num,pb->p->sol[p_order[i]])>=th_dist
     || num==0)
  {
    num++;
    for(j=1;j<=pb->nvar;j++)
      pb->rs->sol[num][j]=pb->p->sol[p_order[i]][j];
    pb->rs->ObjVal[num]=pb->p->ObjVal[p_order[i]];
    pb->rs->HashVal[num]=pb->p->HashVal[p_order[i]];
    pb->rs->iter[num]=1;
  }
}

/* Compute minimum distances from P to RefSet */
min_dist = SSInt_array(pb->p->PSize);
for(i=1;i<=pb->p->PSize;i++)
  min_dist[i]= SSDist_RefSet(pb,b1,pb->p->sol[i]);

/* Add the diverse b-b1 solutions to RefSet */
for(i=b1+1;i<=pb->rs->b;i++)
{
  a=SSMax_dist_index(pb,min_dist);

  for(j=1;j<=pb->nvar;j++)
    pb->rs->sol[i][j] = pb->p->sol[a][j];
  pb->rs->ObjVal[i] = pb->p->ObjVal[a];
  pb->rs->HashVal[i] = pb->p->HashVal[a];
  pb->rs->iter[i]    = 1;

  SSUpdate_distances(pb,min_dist,i);
}

/* Reorder RefSet */
rs_order = SSOrder_i(pb->rs->ObjVal,pb->rs->b,1);
for(i=1;i<=pb->rs->b;i++)
  pb->rs->order[i] = rs_order[i];

pb->rs->NewSolutions = 1;
pb->CurrentIter      = 1;
```

```
        free(rs_order+1);free(p_order+1);free(min_dist+1);
}
```

The SSCreate_RefSet function in file SSRefSet3d.c constructs the reference set. As in Function 2-4, the solutions in *P* are ordered according to their objective function value and the best solutions are deleted from *P* and added to *RefSet*. When adding solutions, the solution values sol, the objective function value ObjVal, and the hash value HashVal are copied from p to rs. (For an explanation of the role of HashVal, see Section 1.3.2 of this chapter.) The main difference between Function 2-4 and Function 5-4 is that the latter first calls function SSDist_RefSet to compute the distance between each selected solution from *P* and those already in the *RefSet*. If the distance is greater than th_dist the solution is added to *RefSet*; otherwise the solution is discarded. In Function 5-4 the th_dist variable has been set to 10. Note that a large value of th_dist could result in a situation where *P* would not be able to yield b_1 solutions that are dispersed enough to be admitted to *RefSet*. The code in Function 5-4 does not provide a way to recover from this situation and the SSAbort function is called to abort the execution of this function and stop the program. The insertion of the $b - b_1$ diverse solutions is made in the same way as in Function 2-4. The files associated with the functions in Section 1.3 of this chapter are in the Chapter5\Section 1.3 folder.

1.3.2   Hashing

Scatter search does not allow duplications in the reference set, because combination methods that operate on reference solutions expect that the solutions in each subset are different from each other. Hashing may be used to reduce the computational effort of checking for duplicated solutions in the reference set. The following hash function, for instance, is an efficient way of comparing solutions and avoiding duplications when dealing with problems whose solutions can be represented with a permutations *p* of size *m*:

$$hash(p) = \sum_{i=1}^{m} i p(i)^2$$

In the context of the linear ordering problem, Campos, et al. (2001) empirically determined that solutions that are different and that share the same objective function value almost always have different *hash* values (with a few rare exceptions). Therefore, when two solutions with the same objective function value have the same hash value, they could either be considered identical or a full duplication checking mechanism could be applied. Campos, et al. (2001) report computational savings of up to 7% when using hashing over a full checking procedure when comparing two solutions that have the same objective function value.

The SSEqualSol function (see Function 5-5) implements a tri-level duplication check to compare sol1 with sol2. First, the objective function values obj1 and obj2 corresponding to sol1 and sol2 are compared. If they are different, a 0 code is returned indicating that the solutions are different. Otherwise, the hashing checking is applied, comparing the corresponding hash values hash1 and hash2. If they are different, a 0 code is returned; otherwise a full duplication check is applied where the value of each variable in both solutions is compared.

Function 5-5. SSEqualSol — File: SSTools3d.c

```
int  SSEqualSol(int  sol1[],int  obj1,int  hash1,int  sol2[],int
obj2,int hash2,int dim)
{
    int i;

    /* Objective value check */
    if(obj1 != obj2)
       return 0;

    /* Hash value check */
    if(hash1 != hash2)
       return 0;

    /* Full duplication check */
    for(i=1;i<=dim;i++)
       if(sol1[i] != sol2[i])
          return 0;

    return 1;
}
```

Function 5-5 assumes that the objective function value is an integer variable. Hence, the comparison obj1 != obj2 is sufficient to determine whether two objective function values are different. When dealing with objective function values in a continuous space, the comparison should be made with reference to a tolerance value. For instance, if the tolerance is specified by the constant OBJTOL, then the expression that determines whether `obj1` is different from `obj2` is `fabs(obj1-obj2) > OBJTOL`. The same applies to the values of `sol1` and `sol2`. For more on hashing in the context of metaheuristic search, we recommend reading Woodruff and Zemel (1993).

2. SUBSET GENERATION

Solution Combination Methods in scatter search typically are not limited to combining just two solutions and therefore the Subset Generation Method in its more general form consists of creating subsets of different sizes. The scatter search methodology is such that the set of combined solutions (i.e., the set of all combined solutions that the implementation intends to generate) may be produced in its entirety at the point where the subsets of reference solutions are created. Therefore, once a given subset is created, there is no merit in creating it again. This creates a situation that differs noticeably from those considered in the context of genetic algorithms, where the combinations are typically determined by the spin of a roulette wheel, as discussed in Section 1 of Chapter 7.

The procedure for generating subsets of reference solutions uses a strategy to expand pairs into subsets of larger size while controlling the total number of subsets to be generated. In other words, the mechanism does not attempt to create all the subsets of size 2, then all the subsets of size 3, and so on until reaching the subsets of size b-1 and finally the entire *RefSet*. This approach would not be practical because there are 1013 subsets in a reference set of a typical size $b = 10$. Even for a smaller reference set, combining all possible subsets is not effective because many subsets will be almost identical. For example, a subset of size four containing solutions 1, 2, 3, and 4 is almost the same as all the subsets with four solutions for which the first three solutions are solutions 1, 2 and 3. And even if the combination of subset {1, 2, 3, 5} were to generate a different solution than the combination of subset {1, 2, 3, 6}, these new trial solutions would likely converge to the same local optimum after the application of the Improvement Method.

The following approach selects representative subsets of different sizes by creating subset types:

- *Subset Type* 1: all 2-element subsets.
- *Subset Type* 2: 3-element subsets derived from the 2-element subsets by augmenting each 2-element subset to include the best solution not in this subset.

- *Subset Type* 3: 4-element subsets derived from the 3-element subsets by augmenting each 3-element subset to include the best solutions not in this subset.
- *Subset Type* 4: the subsets consisting of the best *i* elements, for *i* = 5 to *b*.

A central consideration of this design is that *RefSet* itself might not be static, because it might be changing as new solutions are added to replace old ones (when these new solutions qualify to be among the current *b* best solutions found). Function 5-6 in folder Chapter5\Section 2\Static shows the general form of the Subset Generation Method when the static updating of *RefSet* is utilized. The version that considers the dynamic updating of the reference set along with the generation of subset types 1 to 4 is in folder Chapter5\Section 2\Dynamic.

Function 5-6. SSCombine_RefSet — File: SSRefSet3e.c

```
void SSCombine_RefSet(SS *pb)
{
    int i,j,a,b,c,d,s,onenewsol;
    int *newsol,*sols,*value;

    newsol = SSInt_array(pb->nvar);
    sols   = SSInt_array(pb->rs->b);
    value  = SSInt_array(pb->rs->b);

    /* Combine elements in RefSet */
    for(i=1;i<pb->rs->b;i++)
    for(j=i+1;j<=pb->rs->b;j++)
    {
      /****** SubsetType 1 ************/
      a=pb->rs->order[i];
      b=pb->rs->order[j];
      sols[1]=a;
      sols[2]=b;
      value[1]=pb->rs->ObjVal[a];
      value[2]=pb->rs->ObjVal[b];

      if(pb->rs->iter[a] == pb->CurrentIter ||
         pb->rs->iter[b] == pb->CurrentIter)
      {
        SSCombine_nsol(pb,2,sols,value,newsol);
        pb->pool_size++;
        for(s=1;s<=pb->nvar;s++)
```

```
        pb->pool[pb->pool_size][s]=newsol[s];
  }

  if(j+1<=pb->rs->b)
  {
    /****** SubsetType 2 ************/
    c=pb->rs->order[j+1];
    sols[3]=c;
    value[3]=pb->rs->ObjVal[c];

    if(pb->rs->iter[a] == pb->CurrentIter ||
       pb->rs->iter[b] == pb->CurrentIter ||
       pb->rs->iter[c] == pb->CurrentIter)
    {
      SSCombine_nsol(pb,3,sols,value,newsol);
      pb->pool_size++;
      for(s=1;s<=pb->nvar;s++)
        pb->pool[pb->pool_size][s]=newsol[s];
    }

    if(j+2<=pb->rs->b)
    {
      /****** SubsetType 3 ************/
      d=pb->rs->order[j+2];
      sols[4]=d;
      value[4]=pb->rs->ObjVal[d];

      if(pb->rs->iter[a] == pb->CurrentIter ||
         pb->rs->iter[b] == pb->CurrentIter ||
         pb->rs->iter[c] == pb->CurrentIter ||
         pb->rs->iter[d] == pb->CurrentIter)
      {
        SSCombine_nsol(pb,4,sols,value,newsol);
        pb->pool_size++;
        for(s=1;s<=pb->nvar;s++)
          pb->pool[pb->pool_size][s]=newsol[s];
} } } }

/******** SubsetType 4 ************/
onenewsol=0;
for(i=1;i<=pb->rs->b;i++)
{
```

```
        a=pb->rs->order[i];
        sols[i]=a;
        value[i]=pb->rs->ObjVal[a];
        if(pb->rs->iter[a] == pb->CurrentIter)
          onenewsol=1;

        if(i>=5 && onenewsol)
        {
          SSCombine_nsol(pb,i,sols,value,newsol);
          pb->pool_size++;
          for(s=1;s<=pb->nvar;s++)
            pb->pool[pb->pool_size][s]=newsol[s];
        }
      }

      free(newsol+1);free(sols+1);free(value+1);
}
```

The SSCombine_RefSet function starts with the generation of the subset type 1 elements. The solutions in *RefSet* are scanned according to the order established in the order array. If at least one of the elements in the subset is new, that is, they have not be previously combined (pb->rs->iter = pb->CurrentIter), then the subset is submitted to the Combination Method. The SSCombine_nsol function implements a generalized combination procedure. The procedure is a generalization of the one implemented in Function 4-4 and combines 2 or more reference solutions.

For each type 1 subset with two solutions, one subset of type 2 is derived by adding the next solution in the order array. The array sols stores the indexes of the solutions in the subset and the value array their corresponding objective function values. If at least one of the three elements in a type 2 subset is new, the subset is submitted to the SSCombine_nsol function for combination. Similarly, the type 3 subsets are derived from type 2 subsets and combined. Finally, the type 4 subsets with 5 or more solutions are constructed and combined when appropriate (i.e., when the subsets contains at least one new solutions). All the generated solutions are added to the *Pool* and are processed by the SSUpdate_RefSet function, which tests for possible inclusion in the *RefSet*.

Campos, et al. (2001) designed an experiment with the goal of assessing the contribution of combining subset types 1 to 4 in the context of the linear ordering problem. The experiment undertook to identify how often, across a set of benchmark problems, the best solutions came from combinations of reference solution subsets of various sizes.

Since subset types 1 through 4 respectively generate solutions from 2 to up to *b* reference solutions, it is sufficient to keep a 4 element array for each solution generated during the search. The first element of the array is the counter corresponding to subset type 1; the second element corresponds to subset type 2, and so on. The array for each solution in the initial reference set starts as (0,0,0,0), meaning that there are no sources. The array then counts the number of times the different subset types are used. For instance, suppose three solutions in a subset of type 2 with arrays (2,0,0,1), (5,1,0,0) and (0,1,0,0) are combined. Then a new solution resulting from this combination is accompanied by the array (7,3,0,1)—the sum of the other arrays, plus 1 added to position 2.

In an experiment with 15 instances from LOLIB (see Section 3 in Chapter 4), Campos, et al. (2001) used the tracking arrays to find the percentage of time that each subset type produces solutions that become members of the reference set. They also performed the same experiment employing 15 randomly generated instances with 100 sectors and with weight values uniformly distributed between 0 and 100. The percentages are shown in Figure 5-5, where the bars labeled LOLIB represent the results from the experiments with the LOLIB instances and the bars labeled Random correspond to the results from the randomly generated instances.

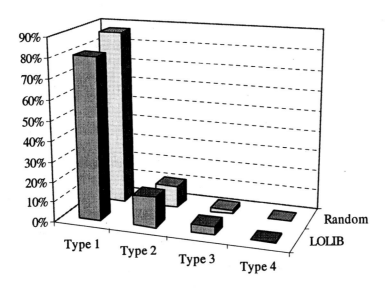

Figure 5-5. Relative merit of subset types

This experiment provides an idea of the relative importance of the various subset types. Although most of the contribution seems to come from subset type 1, these percentages could change if the subset types were generated in a different sequence. Nonetheless, the results of this experiment indicate that a basic SS implementation that employs only subsets of type 1 can be quite effective.

3.   SPECIALIZED COMBINATION METHODS

In the tutorial chapters, we introduced three types of combination methods:

1. Linear combinations (for nonlinear optimization)
2. Score-based combinations (for 0-1 knapsack problems)
3. Combinations by votes (for linear ordering problems)

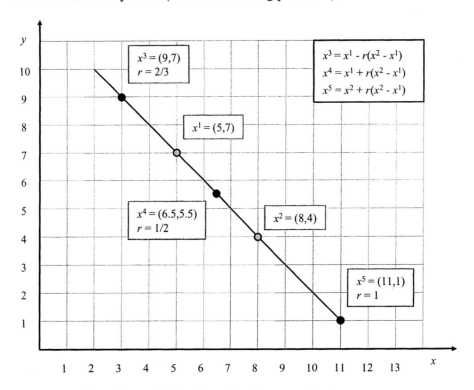

Figure 5-6. Two-dimensional linear combination

Linear combinations are the most intuitive way of combining solutions that are represented by continuous variables. In a two dimensional space, the linear combinations introduced in Chapter 2 are depicted in Figure 5-6. In this figure, x^1 and x^2 are the reference solutions, from which x^3, x^4 and x^5 are generated. Since the same value of r is used for calculating both coordinates of each new solution, the Combination Method consists of generating new trial solutions by sampling the line defined by x^1 and x^2. An alternative design generates one r value for each coordinate of the new trial solution under construction. For instance, suppose that x^4 is being constructed from the combination of x^1 and x^2, as shown in Figure 5-6. However, instead of a single r value for both coordinates of x^4, we use $r = 1/3$ for the x coordinate and $r = 2/3$ for the y coordinate. The new trial solution is $x^4 = (6,5)$ by the following calculations:

$$x = 5 + (1/3)(8 - 5) = 6$$
$$y = 7 + (2/3)(4 - 7) = 5$$

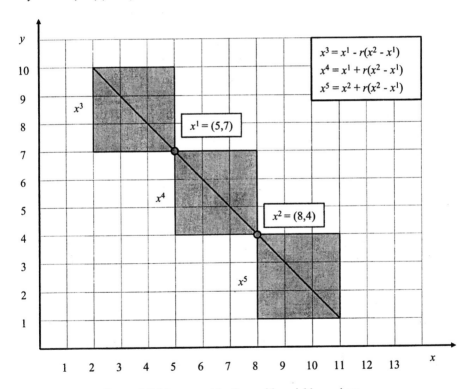

Figure 5-7. Linear combinations with variable r values

In general, if different r values are used for each coordinate and r can take on values between 0 and 1, the Combination Method is capable of sampling the square defined by (5, 4), (8,4), (8,7) and (5,7) when using the formula for x^4. The areas sampled using all three formulas for solutions x^3,

x^4 and x^5 are depicted in Figure 5-7. Experimentation with the linear combination variants depicted in Figures 5-6 and 5-7 is needed to determine their merit in a specific problem setting.

The Combination Method based on scores described in Chapter 3 in the context of 0-1 knapsack problems also has two implementation variants that result from the use of the random variable r. The Combination Method in Chapter 3 uses a uniform random number r ($0 \leq r \leq 1$) to determine whether a variable should be set to zero or one according to the following formula:

$$x_i = \begin{cases} 1 & \text{if } r \leq score(i) \\ 0 & \text{if } r > score(i) \end{cases}$$

In Chapter 3, a single value of r was used for all the variables in a single solution. This is the equivalent of the linear combination shown in Figure 5-6. Alternatively, a different value of r could be used for each variable of a new trial solution. Doing this is equivalent to a linear combination procedure that samples solutions in rectangles instead of being limited to the line joining the reference set, as depicted in Figure 5-7. Here again, computational experiments are necessary to determine the merit of each approach.

3.1   Variable Number of Solutions

In the tutorial chapters, we used a design of the Combination Method that generated the same number of solutions each time any two reference solutions were combined. Specifically, the combination of two reference solutions generated 3, 2 and 1 new trial solutions in the nonlinear optimization example, 0-1 knapsack problems and the linear ordering problem, respectively. While this design might work well in a variety of settings, there exists experimental evidence that not all combinations are likely to produce high quality solutions (Campos, et al. 2001).

Knowing not only how often the best solutions come from different types of combinations, but also knowing the ranks of the reference solutions that generated these best solutions is an issue of particular relevance in scatter search. In other words, it is enlightening, for example, to know that the overall best solution came from combining the 3rd and 5th reference solutions — where one of these came from combining the 1st, 2nd and 6th reference solutions, and the other came from ... etc. This information gives us an idea of the relative importance of the solutions in the reference set with regard to their power of generating new trial solutions of the highest quality.

Experimental evidence can be collected by creating a $b \times b$ matrix *Source*, where *Source*(i,j) counts the number of times a solution of rank j was a reference solution for the current solution with rank i. The rank of a reference solution is given by its position in *RefSet*. Since the reference set is ordered, the best solution has rank 1 and the worst has rank b. To begin, all the elements of *Source* are zero. Then, suppose that a solution to be assigned rank 2 was created from 3 reference solutions, of rank 1, 4 and 6. Before shifting the records so that the previous rank 2 will be the new rank 3, etc., we create a "working space" *Work(j)* that will become the row *Source(2,j)* after the shift. *Work(j)* in the present example just sums the contents of the three rows *Source(1,j)*, *Source(4,j)* and *Source(6,j)*, and then also increases *Work(1)*, *Work(4)* and *Work(6)* by 1. The data structure is maintained and updated by this basic type of process.

Figure 5-8 shows the average of rows 1, 10 and 20 of the *Source* matrix after running scatter search on 15 LOLIB instances. The interpretation of the lines in Figure 5-8 is as follows. Consider the line associated with rank 1. Then, the count of (almost) 18 in the first point of this line indicates that rank 1 solutions were generated (approximately) 18 times from other rank 1 solutions. Similarly, rank 1 solutions were generated 8 times from rank 2 solutions. The decaying effect exhibited by all the lines indicates that high quality solutions tend to generate new solutions that are admitted to the reference set. This is evident by the counts corresponding to rank 1 in the horizontal axis of Figure 5-8. In this experiment, rank 1 solutions generated 18 rank 1 solutions, 4 rank 10 solutions and 1 rank 20 solution, on average.

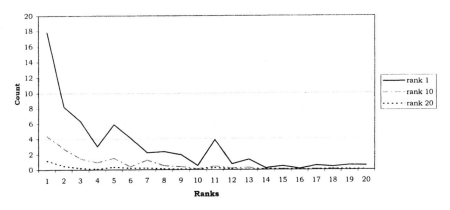

Figure 5-8. Plot of rows 1, 10 and 20 from the *Source* matrix

Since there is evidence that high-ranked solutions are more likely to generate high-quality solutions that become reference solutions, it seems sensible that a specialized Combination Method should not treat all combinations equally. For instance, the combination of solutions 1 and 2 in the reference set should be treated differently than the combination of solutions b-1 and b. A common rule that exploits this information uses the parameters b_1 and b_2, such that $b = b_1 + b_2$, to determine the number of solutions to generate from the combination of the reference solutions x^i and x^j. The number of new trial solutions, n, is set as follows:

- $n = n_1$ if $i \leq b_1$ and $j \leq b_1$
- $n = n_2$ if $i \leq b_1$ and $j \geq b_1$
- $n = n_3$ if $i \geq b_1$ and $j \geq b_1$

To preserve the aggressive nature of scatter search implementations, the relationship among n_1, n_2 and n_3 should be such that $n_1 > n_2 > n_3$. Other rules are possible that involve the use of the b_1 and b_2 parameters for reference sets that are updated as done in the basic scatter search scheme. Additional rules can also be created for the reference set updates discussed in Sections 1.2.2 and 1.2.3 of this chapter, which partition *RefSet* into *RefSet*$_1$, *RefSet*$_2$ and *RefSet*$_3$.

3.2   Binary Variables

When generating linear combinations of solutions it is important to be sure the results are dispersed appropriately, and to avoid calculations that simply duplicate other — particularly where the duplications occur after mapping continuous values into integer values. This caveat is especially to be heeded in the case of zero-one variables, since each continuous value can only map into one of the two alternatives 0 and 1, and a great deal of effort can be wasted by disregarding this obvious consequence.

Specifically, for the 0-1 case, a useful set of combinations and mappings (except for random variations) of r reference solutions consists simply of those that create a positive integer threshold $t \leq r$, and require that the new trial solution will have its ith component $x_i = 1$ if and only if at least t of the r reference solutions being combined have $x_i = 1$. A preferred way to apply this rule is to subdivide the variables into categories, and to apply different thresholds to different categories—i.e., to different subvectors. Many problems offer natural criteria for subdividing variables into categories. However, in the absence of this, or in the case where greater diversity is desired, the categories can be arbitrarily generated and varied. At the extreme, for example, each variable can belong to its own category. This scatter search option gives the same set of alternative mappings as the so-called *Bernoulli crossover* introduced into genetic algorithms about a decade after the original scatter search proposals, except that scatter search does not select the 0 and 1 values at random as in the Bernoulli scheme, but instead selects these values by relying on frequency memory to achieve intensification and diversification goals, as discussed in Chapter 6.

To illustrate the consequences of such a threshold mechanism applied to 2 reference solutions (or to selected subvectors of two reference solutions), $t = 1$ gives the new trial solution that is the union of the 1's in the reference solutions and $t = 2$ gives the new trial solution that is the intersection of the 1's in the reference solutions. But the threshold mechanism is not the only relevant one to consider. Two other kinds of combinations can be produced for $r = 2$, consisting of those equivalent to subtracting one vector (or subvector) from another and setting $x_i = 1$ if and only if the difference is positive. In this case it is not unreasonable to consider one more linear combination by summing these two new trial solutions to get the symmetric difference between the reference solutions.

To summarize and restricting attention to $r = 2$ yields the following options of interest:

1. the intersection of 1's
2. the union of 1's
3. the 1's that belong to reference solution 1 and not to reference solution 2
4. the 1's that belong to reference solution 2 and not to reference solution 1
5. the symmetric difference (summing (3) and (4)).

In some contexts the last three options are not as useful as the first two, but in others they can be exceedingly important. An example of the importance of the symmetric difference is given in Glover and Laguna (1997). An analogous set of options for $r = 3$ is given by:

1. the intersection of 1's
2. the union of 1's
3. the 1's that belong to a majority of the reference solutions
4. the 1's that belong to exactly 1 reference solution but not more
5. the 1's that belong to exactly 2 reference solutions but not more
6. the 1's that belong to the union of 2 reference solutions, excluding those that belong to the 3rd reference solution. (3 different cases)
7. the 1's that belong to the intersection of 2 reference solutions, excluding those that belong to the 3rd reference solutions. (3 different cases)

Inclusion of the 6th and 7th options, each of which involves 3 different cases, entails a fair amount of effort, which may not be warranted in many situations. However, it would make sense to use (6) and (7) by defining the "3rd reference solution" to be the worst of the 3, hence the 1's that are excluded are those that belong to the worst reference solution.

In general, the relevant mappings for linear combinations involving 0-1 variables can be identified in advance by rules such as those indicated, applied either to full vectors or to subvectors. These results is a much more economical and effective process than separately generating a wide range of linear combinations and then performing a mapping operation. The latter can produce multiple duplications and even miss relevant alternatives.

4.   DIVERSIFICATION GENERATION

Specialized Diversification Generation Methods are often customized to specific problem settings. For instance in Chapter 2, we described a method based on frequency counts that was customized to nonlinear optimization problems with continuous variables. This method uses frequency data and controlled randomization to create diversified solutions to build or rebuild the reference set. In Chapter 3 we took a different approach by designing a systematic method for generating diversification. Although limited in some ways, the method in Chapter 3 is a good example of the use of problem structure and solution representation to create a method that does not rely on randomization to achieve diversity. Taking advantage of other methodologies that have their own framework for generating diversity is another alternative, as we showed in Chapter 4.

In this section, we expand upon the methods described in the tutorial chapters. First we take a look at the use of experimental design in the context of problems with solutions represented by continuous variables. We then describe additional diversification generators based on GRASP constructions, which expands upon the material presented in Chapter 4.

4.1   Experimental Design

The basic Diversification Generation Method in Chapter 2 divides the range of each variable $u_i - l_i$ into 4 sub-ranges of equal size. Then, a solution is constructed in two steps. First a sub-range is randomly selected. The probability of selecting a sub-range is inversely proportional to its frequency count. Then a value is randomly generated within the selected sub-range. The number of times sub-range j has been chosen to generate a value for variable i is accumulated in $freq(i, j)$.

Two variants of this method are based on techniques from the area of statistics known as Design of Experiments. One of the most popular design of experiments is the factorial design k^n, where n is the number of factors (in our case variables) and k is the number of levels (in our case feasible variable values). A full factorial design considers that all combinations of the factors and levels will be tested. Therefore, a full factorial design with 5 factors and 3 critical levels would require $3^5 = 243$ experiments. Obviously, full factorial designs can quickly become impractical even for a small number of levels, because the number of experiments exponentially increases with the number of factors. A more practical alternative is to employ fractional factorial designs. These designs draw conclusions based on a fraction of the experiments, which are strategically selected from the set of all possible experiments in the corresponding full factorial design. One of the most notable proponents of the fractional factorial designs is Genichi Taguchi (Roy, 1990). Taguchi proposed a special set of orthogonal arrays to lay out experiments associated with quality improvement in manufacturing. These orthogonal arrays are the result of combining orthogonal Latin squares in a unique manner. We propose applying Taguchi's arrays as a mechanism for generating diversity. Table 5-2 shows the $L_9(3^4)$ orthogonal array that would generate 9 solutions for a 4-variable problem.

Table 5-2. $L_9(3^4)$ orthogonal array

Experiment	Factors			
	1	2	3	4
1	1	1	1	1
2	1	2	2	2
3	1	3	3	3
4	2	1	2	3
5	2	2	3	1
6	2	3	1	2
7	3	1	2	3
8	3	2	1	3
9	3	3	2	1

The values in Table 5-2 represent the levels at which the factors are set in each experiment. For the purpose of creating a Diversification Generation Method based on Taguchi tables, we translate each level setting as follows:

$1 = $ lower bound (l_i)

$2 = $ midpoint $\left(\dfrac{l_i + u_i}{2} \right)$

$3 = $ upper bound (u_i)

The procedure access the appropriate Taguchi table to draw solutions every time the Diversification Generation Method is called. The Taguchi table is selected according to the number of variables in the problem. Because the number of rows in the Taguchi tables is a linear function of the number of variables, the diversity generation could revert to the basic pseudo-random procedure once all solutions from the appropriate table have been tried.[1]

A second variant of a Diversification generation Method based on Taguchi tables is obtained when level settings in each experiment are translated to variable values as follows:

$1 =$ near lower bound $(l_i + r(u_i - l_i))$, where r is randomly drawn from $(0, 0.1)$

$2 =$ near midpoint $\left(\dfrac{l_i + u_i}{2} + r(u_i - l_i)\right)$, where r is randomly drawn from $(-0.1, 0.1)$

$3 =$ near upper bound $(u_i - r(u_i - l_i))$, where r is randomly drawn from $(0, 0.1)$

Although the same Taguchi experiment in this case can result in more than one solution to the problem, it is preferable to turn control to the basic Diversification Generation Method once all experiments have been used to generate solutions, because the basic method covers the entire feasible range of each variable, while the second variant based on Taguchi tables focuses on generating solutions with variable values near the bounds and the midpoint.

4.2   GRASP Constructions

In Chapter 4, we provided a general description and a high-level view of the GRASP methodology.   We then showed how the adaptive and constructive elements of the methodology could be applied to the generation of diverse solutions to the linear ordering problem. Particularly, we obtained GRASP constructions with the attractiveness of a sector defined by the sum of the elements in its corresponding row (DG01 method).   The method randomly selected from a short list of the most attractive sectors and constructed the solution starting from the first position of the permutation.

Additional methods based on GRASP constructions can be obtained by changing some operational rules. For example:

---

[1]   Taguchi tables are available in http://www.york.ac.uk/depts/maths/tables/orthogonal.htm and additional orthogonal arrays in http://www.research.att.com/~njas/oadir/.

- The measure of attractiveness of a sector may be the sum of the elements in its corresponding column. (DG02 method)
- The measure of attractiveness of a sector may be the sum of the elements in its corresponding row divided by the sum of the elements in its corresponding column. (DG03 method)
- Variants of all the previous methods where the selection of sectors is from a short list of the least attractive, and the solution is constructed starting from the last position of the permutation. (DG04, DG05 and DG06 methods)
- A mixed procedure derived from the previous 6. The procedure generates the same number of solutions from each of the previous six methods until reaching the desired total number of solutions in a way that each method contributes with the same number of solutions. (DG07 method)

Campos, et al. (2001) developed and tested three additional methods for generating diversity in the context of the linear ordering problem. One of them was a pure random procedure (DG08) and another completely deterministic (DG09). Finally, a method based on frequency memory, which we describe in Chapter 6, was also included in the set of diversification generators (DG10). In order to determine the most effective method, two measures where created: one of diversity (Δd) and one of quality (ΔC). Figure 5-9 shows the Δd, ΔC and $\Delta C + \Delta d$ values achieved by each procedure.

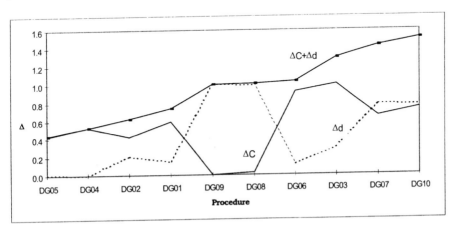

Figure 5-9. Diversity vs. quality for alternative Diversification Generation Methods

Since both the quality and diversity measures are standardized in such a way that $0 \le \Delta d \le 1$ and $0 \le \Delta C \le 1$, the maximum score that a procedure can obtain is 2 and the minimum is 0. As expected, the random generator (DG08) produces the maximum diversity (as measured by Δd), but at the same time produces solutions of the lowest quality. DG09 matches the diversity of DG08 using a systematic approach instead of randomness, but also fails in producing solutions of a reasonable quality. The mixed method DG07 provides a good balance between diversity and quality, by the union of solutions generated with methods DG01 to DG06. Clearly, DG10 outperforms the competing methods, by producing the highest combined score and an almost perfect balance between diversification and solution quality. Experiments such as the one summarized in Figure 5-9 are quite valuable when considering several variants of Diversification Generation Methods.

Chapter 6

USE OF MEMORY IN SCATTER SEARCH

Never rat on your friends and always keep your mouth shut.

Robert de Niro in Goodfellas (1990)

Evolutionary approaches, such as scatter search, implicitly make use of memory. In fact, the following quote from Glover and Laguna (1997) argues that in a sense all heuristic procedures make use of memory in one way or another:

From a naive standpoint, virtually all heuristics other than complete randomization induce a pattern whose present state depends on the sequence of past states, and therefore incorporate an implicit form of "memory." Given that the present is inherited from the past, the accumulation of previous choices is in a loose sense "remembered" by current choices. This sense is slightly more pronounced in the case of solution combination methods such as genetic algorithms and scatter search, where the mode of combination more clearly lends itself to transmitting features of selected past solutions to current solutions. Such an implicit memory, however, does not take a form normally viewed to be a hallmark of an intelligent memory construction. In particular, it uses no conscious design for recording the past and no purposeful manner of comparing previous states or transactions to those currently contemplated. By contrast, at an opposite end of the spectrum, procedures such as branch and bound and A* search use highly (and rigidly) structured forms of memory — forms that are organized to generate all nondominated solution alternatives with little or no duplication.

The implicit use of memory in scatter search is evident if one examines how the Reference Set Update, Solution Combination and Subset Generation Methods operate. The Reference Set Update Method, in its most basic form, is designed to "remember" the best solutions encountered during the search. Some features of these solutions are used to create new trial solutions with the Combination Method. Hence, this method is instrumental in the transmission of information embedded in the reference solutions.

Maintaining a set of solutions throughout the search and combining these solutions to generate new ones does not make scatter search unique regarding the implicit use of memory within an evolutionary framework. Genetic algorithms also operate on a population of solutions and the "survivable of the fittest" philosophy translates into an implicit use of memory. The Subset Generation Method, however, makes even a basic scatter search implementation a more "memory oriented" procedure than a genetic algorithm. From one iteration to the next, this method keeps track of the subsets of reference solutions that have already been subjected to the combination mechanism. When new solutions are admitted to the reference set, the method generates only those subsets that are admissible for combination in the current iteration. The Subset Generation Method performs this operation using a memory structure that allows it to identify the subsets that contain new reference solutions. There is no equivalent use of memory in genetic algorithms, since this approach selects solutions for combination purposes using a random sampling scheme as described in Section 1 of Chapter 7.

Tabu search is perhaps the metaheuristic procedure that employs memory in the most strategic and direct way. Tabu search and scatter search have a common history; their basic principles were suggested by Glover (1977). In Section 1.2 of Chapter 1 we describe a hybrid method for combining the evolutionary features of scatter search with the memory-driven strategies of tabu search. Before we expand upon such design and discuss other uses of memory within scatter search, we provide a brief description of tabu search for the benefit of those readers not familiar with this metaheuristic. Our description is a synthesis of material appearing in Glover and Laguna (1997).

1.   TABU SEARCH

Tabu search (TS) is based on the premise that problem solving, in order to qualify as intelligent, must incorporate *adaptive memory* and *responsive exploration*. The adaptive memory feature of TS allows the implementation of procedures that are capable of searching the solution space economically and effectively. Since local choices are guided by information collected during the search, TS contrasts with memoryless designs that heavily rely on semirandom processes that implement a form of sampling.

The emphasis on responsive exploration in tabu search, whether in a deterministic or probabilistic implementation, derives from the supposition that a bad strategic choice can yield more information than a good random choice. In a system that uses memory, a bad choice based on strategy can provide useful clues about how the strategy may profitably be changed. (Even in a space with significant randomness a purposeful design can be more adept at uncovering the imprint of structure.)

The memory used in tabu search is both *explicit* and *attributive*. Explicit memory records complete solutions, typically consisting of elite solutions visited during the search. An extension of this memory records highly attractive but unexplored neighbors of elite solutions. The memorized elite solutions (or their attractive neighbors) are used to expand the local search.

TS uses attributive memory for guiding purposes. This type of memory records information about solution attributes that change in moving from one solution to another. For example, in a graph or network setting, attributes can consist of nodes or arcs that are added, dropped or repositioned by the moving mechanism. In production scheduling, the index of jobs may be used as attributes to inhibit or encourage the method to follow certain search directions.

Two highly important components of tabu search are intensification and diversification strategies. Intensification strategies are based on modifying choice rules to encourage move combinations and solution features historically found good. They may also initiate a return to attractive regions to search them more thoroughly. Since elite solutions must be recorded in order to examine their immediate neighborhoods, explicit memory is closely related to the implementation of intensification strategies.

Tabu search begins in the same way as ordinary local or neighborhood search, proceeding iteratively from one solution to another until a chosen termination criterion is satisfied. When TS is applied to an optimization problem with the objective of minimizing or maximizing $f(x)$ subject to $x \in X$, each x has an associated neighborhood $N(x) \subset X$, and each solution $x' \in N(x)$ is reached from x by an operation called a *move*.

When contrasting TS with a simple descent method where the goal is to minimize $f(x)$ (or a corresponding ascent method where the goal is to maximize $f(x)$), we must point out that such a method only permits moves to neighbor solutions that improve the current objective function value and ends when no improving solutions can be found. Tabu search, on the other hand, permits moves that deteriorate the current objective function value but the moves are chosen from a modified neighborhood $N^*(x)$. Short and long term memory structures are responsible for the specific composition of $N^*(x)$. In other words, the modified neighborhood is the result of maintaining a selective history of the states encountered during the search. A typical tabu search trajectory is shown in Figure 6-1.

The objective function value in the example shown in Figure 6-1 is the weight of arcs in a graph problem (see Chapter 2 of Glover and Laguna, 1997). A constructive greedy heuristic yields an initial solution with a weight of 40 and after the first three nonimproving moves the objective function value of the current solution is 63. The next two moves are improving and the current solution becomes the best solution with a total weight of 37 after iteration 5. The next iteration consists of a move with value equal to zero and therefore the objective function of the current solution remains the same. In iterations 7 and 8 the moves deteriorate the objective function value of the current solution. The best solution, with an objective function value of 34, is found in iteration 9.

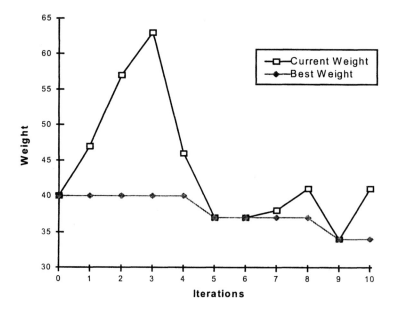

Figure 6-1. Typical tabu search trajectory

In the TS strategies based on short term considerations, $N^*(x)$ characteristically is a subset of $N(x)$, and the tabu classification serves to identify elements of $N(x)$ excluded from $N^*(x)$. In TS strategies that include longer term considerations, $N^*(x)$ may also be expanded to include solutions not ordinarily found in $N(x)$. Characterized in this way, TS may be viewed as a dynamic neighborhood method. This means that the neighborhood of x is not a static set, but rather a set that can change according to the history of the search. Characteristically, a TS process based strictly on short term strategies may allow a solution x to be visited more than once, but it is likely that the corresponding reduced neighborhood $N^*(x)$ will be different each time. With the inclusion of longer term considerations, the likelihood of duplicating a previous neighborhood upon revisiting a solution, and more generally of making choices that repeatedly visit only a limited subset of X, is all but nonexistent.

Recency-based memory is the most common memory structure used in TS implementations. As its name suggests, this memory structure keeps track of solutions attributes that have changed during the recent past. To exploit this memory, selected attributes that occur in solutions recently visited are labeled *tabu-active*, and solutions that contain tabu-active elements, or particular combinations of these attributes, are those that become tabu. This prevents certain solutions from the recent past from belonging to $N^*(x)$ and hence from being revisited. Other solutions that share such tabu-active attributes are also similarly prevented from being visited. Note that while the tabu classification strictly refers to solutions that are forbidden to be visited, by virtue of containing tabu-active attributes (or more generally by violating certain restriction based on these attributes), moves that lead to such solutions are also often referred to as being tabu.

Frequency-based memory provides a type of information that complements the information provided by recency-based memory, broadening the foundation for selecting preferred moves. Like recency, frequency often is weighted or decomposed into subclasses. Also, frequency can be integrated with recency to provide a composite structure for creating penalties and inducements that modify move evaluations.

Frequencies typically consist of ratios, whose numerators represent counts expressed in two different measures: a *transition measure* — the number of iterations where an attribute changes (enters or leaves) the solutions visited on a particular trajectory, and a *residence measure* — the number of iterations where an attribute belongs to solutions visited on a particular trajectory, or the number of instances where an attribute belongs to solutions from a particular subset. The denominators generally represent one of three types of quantities: (1) the total number of occurrences of all events represented by the numerators (such as the total number of associated iterations), (2) the sum (or average) of the numerators, and (3) the maximum numerator value. In cases where the numerators represent weighted counts, some of which may be negative, denominator (3) is expressed as an absolute value and denominator (2) is expressed as a sum of absolute values (possibly shifted by a small constant to avoid a zero denominator). The ratios produce *transition frequencies* that keep track of how often attributes change, and *residence frequencies* that keep track of how often attributes are members of solutions generated. In addition to referring to such frequencies, thresholds based on the numerators alone can be useful for indicating when phases of greater diversification are appropriate.

The use of recency and frequency memory in tabu search creates a balance between search intensification and diversification. Intensification strategies are based on modifying choice rules to encourage move combinations and solution features historically found good. They may also initiate a return to attractive regions to search them more thoroughly. Diversification strategies, on the other hand, increase the effectiveness in exploring the solution space of search methods based on local optimization. Some of these strategies are designed with the chief purpose of preventing searching processes from *cycling,* i.e., from endlessly executing the same sequence of moves (or more generally, from endlessly and exclusively revisiting the same set of solutions). Others are introduced to impart additional robustness or vigor to the search.

2.   EXPLICIT MEMORY

Explicit memory is the approach of storing complete solutions and generally is used in a highly selective manner, because it can consume an enormous amount of space and time when applied to each solution generated. One of the main uses of explicit memory is in the context of optimizing simulations, where scatter search has been applied, as detailed in Chapter 9.

It is widely acknowledged that simulation is a powerful computer-based tool that enables decision-makers in business and industry to improve operating and organizational efficiency. The ability of simulation to model a physical process on the computer, incorporating the uncertainties that are inherent in all real systems, provides an enormous advantage for analysis in situations that are too complex to represent by "textbook" mathematical formulations.

In spite of its acknowledged benefits, however, simulation has suffered a limitation that has prevented it from uncovering the best decisions in critical practical settings. This limitation arises out of an inability to evaluate more than a fraction of the immense range of options available. Practical problems in areas such as telecommunications, manufacturing, marketing, logistics and finance typically pose vast numbers of interconnected alternatives to consider. As a consequence, the decision making goal of identifying and evaluating the best (or near best) options has been impossible to achieve in many applications.

Theoretically, the issue of identifying best options falls within the realm of optimization. Until quite recently, however, the methods available for finding optimal decisions have been unable to cope with the complexities and uncertainties posed by many real world problems of the form treated by simulation. In fact, these complexities and uncertainties are the primary reason that simulation is chosen as a basis for handling such problems. Consequently, decision makers have been faced with the "Catch 22" that many important types of real world optimization problems can only be treated by the use of simulation models, but once these problems are submitted to simulation there are no optimization methods that can adequately cope with them. In short, there has not existed any type of search process capable of effectively integrating simulation and optimization.

When optimizing simulations, decision variables x in the optimization model are the input factors of the simulation model (see Figure 6-2). The simulation model generates an output $f(x)$ for every set of input values. On the basis of this evaluation, and on the basis of the past evaluations which are integrated and analyzed with the present simulation outputs, a scatter search generates a new set of input values.

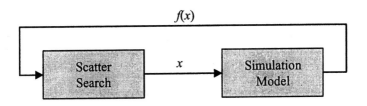

Figure 6-2. Scatter search optimization of a simulation model

Because the evaluation of a set of input values by means of the execution of a simulation model may require a large computational effort, recording every solution generated and evaluated during the search is an effective approach in this setting. When a new trial solution is generated with the Solution Combination Method and before it is sent to the simulation model for evaluation, the explicitly memory is consulted to make sure that the trial solution is indeed new. Since typical optimization runs in this context make at most ten thousand calls to the Simulation Model, the explicit memory structure does not grow to an unmanageable size. Nonetheless, hashing may be used to limit the memory requirements and the effort associated with checking whether a trial solution is already stored in the explicit memory structure. For details of hashing and an example of a hash function, see Section 1.3.2 of Chapter 5.

3.   ATTRIBUTIVE MEMORY

Instead of recording full solutions, attributive memory structures are based on recording attributes. Attributive memory is used for implementing both diversification and intensification strategies. The most common attributive memory approaches are recency-based memory and frequency-based memory. Some examples of recency-based memory are shown in Table 6-1.

Table 6-1. Examples of recency-based memory

Context	Attributes	To record the last time …
Binary problems	Variable index (i)	variable i changed its value from 0 to 1 or 1 to 0 (depending on its current value).
Job sequencing	Job index (j)	job j changed positions.
	Job index (j) and position (p)	job j occupied position p.
	Pair of job indexes (i,j)	job i exchange positions with job j.
Graphs	Arc index (i)	arc i was added to the current solution.
		arc i was dropped from the current solution.

Recency-based memory uses either single or combined attributes to keep track of information that is relevant to the operation of the search procedure, as shown in Table 6-1 for the job sequencing example. Table 6-2 shows examples of frequency-based memory using transition and residence measures.

Table 6-2. Examples of frequency-based memory

Context	Residence measure	Transition measure
Binary problems	Number of times variable i has been assigned the value of 1.	Number of times variable i has changed values.
Job sequencing	Number of times job j has occupied position p.	Number of times job i has exchanged positions with job j.
	Average objective function value when job j occupies position p.	Number of times job j has been moved to an earlier position in the sequence.
Graphs	Number of times arc i has been part of the current solution.	Number of times arc i has been deleted from the current solution when arc j has been added.
	Average objective function value when arc i is part of the solution.	Number of times arc i has been added during improving moves.

The attributes necessary to implement the frequency-based memory shown in Table 6-2 can be easily identified from the context. For example in the context of binary problems, the variable index is the only attribute needed. In the next subsection, we illustrate the use of frequency-based memory for achieving diversification in a scatter search implementation. We then finish the chapter with an illustration of an Improvement Method that employs recency-based memory.

3.1   Diversification

In Section 1 of Chapter 2, we introduced the use of a simple frequency-based memory in connection with a Diversification Generation Method for nonlinear function optimization in a continuous solution space. The range of each variable is divided into four subranges and the method keeps track of the frequency that each subrange is sampled to construct new trial solutions. The attribute in this case is the identity of each subrange and, therefore, in a n-variable problem there are $4*n$ attributes. This frequency-based memory falls within the category of a residence measure by viewing the construction of P as an iterative process that "visits" one solution per iteration for *PSize* iterations.

In Section 4.2 of Chapter 5, we discussed Diversification Generation Methods based on GRASP constructions. In the context of the linear ordering problem, we provided several examples of such constructions. In this section, we provide details of the best method based on GRASP constructions that appears in Figure 5-9 under the label DG10.

The Diversification Generation Method is based on the notion of constructing solutions employing modified frequencies. The generator exploits the permutation structure of a linear ordering. The residence measure employed in this procedure is such that a frequency counter is maintained to record the number of times a sector i appears in position k. The frequency counters are used to penalize the "attractiveness" of a sector with respect to a given position, as in the approach of Laguna and Glover (1993). To illustrate, suppose that the generator has created 30 solutions to be included in P. If 20 of the 30 solutions have sector 3 in position 5, then the frequency counter $freq(3,5) = 20$. This frequency value is used to bias the potential assignment of sector 3 in position 5 during subsequent constructions, and therefore, inducing diversification with respect to the solutions already in P.

The attractiveness of assigning sector i is given by the greedy function $G(i)$, which depends on the set U of unassigned sectors as proposed in Becker (1967).

$$G(i) = \frac{\displaystyle\sum_{j \in U} e_{ij}}{\displaystyle\sum_{j \in U} e_{ji}}$$

As defined in Section 1 of Chapter 4, e_{ij} denotes the amount of deliveries (in monetary value) from sector i to sector j in a given year. We modify the value of $G(i)$ to reflect previous assignments of sector i to position k as follows:

$$G'(i,k) = G(i) - \beta * \left(\frac{MaxG}{MaxF} \right) * freq(i,k)$$

where $MaxF$ is the maximum $freq(i,k)$ value for all i and k, and $MaxG$ is the maximum $G(i)$ value for all $i \in U$.

It is important to point out that $G(i)$ is an adaptive function since its value depends on the weights of the unassigned sectors at each iteration of the construction procedure. A pseudo-code of the Diversification Generation Method to produce a set P with $PSize$ solutions appears in Figure 6-3.

---

1. $P = \emptyset$.
2. $freq(i,j) = 0$, for all i and j.
while $(|P| < PSize)$ **do**
    3. $p(i) = 0$ for all i.
    4. $U = \{ 1,..., m \}$ (The set of unassigned sectors.)
    5. $k = MaxF = MaxG = 1$.
    while $(U \neq \emptyset)$ **do**

      6. $G'(i,k) = G(i) - \beta * \left(\dfrac{MaxG}{MaxF} \right) * freq(i,k)$ $\forall i \in U$

      7. $i^* = \arg \max_i \left(G'(i,j), i \in U \right)$

      8. $MaxG = G(i^*)$
      9. $U = U - \{ i^* \}$
      10. $p(i^*) = k$
      11. $k = k + 1$
    end while
    12. $freq(i,p(i)) = freq(i,p(i)) + 1$, for all i.
    13. $MaxF = \max_{i,k} \left(freq(i,k), i \in U, k \in \{1,...,m\} \right)$
    if $(p \notin P)$ **then** $P = P \cup \{p\}$
end while

Figure 6-3. Diversification Generation Method DG10 for linear ordering problems

The pseudo-code in Figure 6-3 is written using general mathematical notation, where p represents a solution (permutation). However, the actual implementation in the following subsection takes advantage of quick updating mechanisms that are hard to represent mathematically. For example, the value of *MaxF* in step 13 can be maintained by keeping track of the updates of the frequency matrix *freq*. The performance of this Diversification Generation Method as measure by the quality and diversity of the solutions that is capable of generating depends on the value of the β-parameter. Experiments conducted by Campos et al. (2001) determined that the best setting for this parameter should be 0.3.

3.1.1   Computer Code

The following data structures have been added to the ss structure to store all the necessary information for the Diversification Generation Method shown in Figure 6-3:

- The row_sol array stores the sum of the elements in each row.
- The col_sol array stores the sum of the elements in each column.
- The freq matrix stores the frequency counter *freq(i,j)*.
- The Maxfreq variable contains the value of *MaxF*.

Function 6-1, SSGenerate_Sol, is an implementation of the procedure in Figure 6-3 that uses the sol array to return the new trial solution. The function starts with creating a copy of row_sol and col_sol onto the local arrays row_s and sol_s, respectively. Then, at each iteration j, the function computes the weights (attractiveness) as row_s divided by col_s. (Since both variables are integer, a casting to the double type is performed.) Then, the weights are modified according to the frequency values pb->freq as described above.

The unassigned element with maximum modified weight is selected and assigned to sol[j]. Then, the row_s values are modified by subtracting the weights in the sol[j] column. Similarly, the sol_s values are updated by subtracting the weights in the sol[j] row. When a solution has been fully completed after nvar assignments, the frequencies are updated as well as the Maxfreq value.

Function 6-1. SSGenerate_Sol — File: SSP3g.c

```
void SSGenerate_Sol(SS *pb,int sol[])
{
    int i,j,a,*assigned,*row_s,*col_s;
    double *weight,max_weight,beta=0.3;

    assigned = SSInt_array(pb->nvar);
    row_s    = SSInt_array(pb->nvar);
```

```
col_s    = SSInt_array(pb->nvar);
weight   = SSDouble_array(pb->nvar);

/* Initialize row and column sums */
for(i=1;i<=pb->nvar;i++)
{
  row_s[i] = pb->row_sum[i];
  col_s[i] = pb->col_sum[i];
}

/* Iteration j selects element sol[j] */
for(j=1;j<=pb->nvar;j++)
{
  /* Compute (adapt) weights */
  max_weight=-MAXPOSITIVE;
  for(i=1;i<=pb->nvar;i++)
  {
    if(col_s[i]==0) weight[i]=row_s[i];
    else weight[i]=(double)row_s[i]/(double)col_s[i];
    if(!assigned[i] && weight[i]>max_weight)
        max_weight=weight[i];
  }
  for(i=1;i<=pb->nvar;i++)
    weight[i] -=  beta * (max_weight/(double)pb->Maxfreq)
                  *pb->freq[i][j];

  /* Maximum weight of non-assigned sectors */
  max_weight=-MAXPOSITIVE;
  for(i=1;i<=pb->nvar;i++)
    if(!assigned[i] && weight[i]>max_weight)
    {
      a=i;
      max_weight=weight[i];
    }

  /* Assign the selected element */
  assigned[a]=1;
  sol[j]=a;

  /* Adapt weights */
  for(i=1;i<=pb->nvar;i++)
  {
```

```
        row_s[i] -= pb->data[i][a];
        col_s[i] -= pb->data[a][i];
    }
  }


  /* Update frequencies and the maximum */
  for(i=1;i<=pb->nvar;i++)
    if(++pb->freq[sol[i]][i] > pb->Maxfreq )
      pb->Maxfreq = pb->freq[sol[i]][i];


  free(row_s+1);free(col_s+1);
  free(weight+1);free(assigned+1);
}
```

3.2   Intensification

Search intensification is typically achieved in scatter search with the execution of the Improvement Method. Basic Improvement Methods are based on implementations of local search procedures. For instance, we used the well-known Nelder and Mead procedure for search intensification purposes in Chapter 2. In Chapter 3 and 4, we described local search procedures that we adapted for each situation: knapsack and linear ordering problems. None of these procedures, however, used memory or explored beyond the first local optimum found during the search.

The use of memory in connection with the Improvement Method raises some interesting issues regarding the computational effort associated with each of the methods in a scatter search implementation. An Improvement Method consisting of a pure local search has a natural stopping rule and the computational effort devoted to improving solutions is controlled with rules that determine whether or not to apply the method to all or only a subset of the solutions generated during the search. When memory structures are added to the Improvement Method, the method is conceptually and practically transformed from a pure local search heuristic into a metaheuristic given that by definition a metaheuristic refers to a master strategy that guides and modifies other heuristics to produce solutions beyond those that are normally generated in a quest for local optimality.

Employing a metaheuristic as an Improvement Method results in a hybrid method that combines two metaheuristics, i.e., scatter search and the one used to improve solutions. An important issue in such a design is how to allocate the total computational effort. In other words, should the search spend most of the time improving solutions or generating new trial solutions with the Diversification Generation and Combination Methods?

Metaheuristic procedures don't have a natural stopping criterion, because they are designed to keep exploring the solution space with the goal of finding better local optimal points. Hence, the balance in the computational effort must be controlled by not only selecting the solutions to be subjected to the Improvement Method but also by choosing a rule to stop the improvement process.

As an illustration of using memory within an Improvement Method, let us modify the Improvement Method for the linear ordering problem described in Section 3 of Chapter 4. The resulting method is a so-called short-term memory tabu search. Instead of examining the neighborhood of each sector in descending order of their weight, in each iteration a sector is probabilistically selected. The probability of selecting sector j is proportional to its weight w_j. That is, the larger the weight, the larger the probability of being selected. The move INSERT_MOVE($p(j)$, i) $\in N_2^j$ with the largest move value is selected. (Note that this rule may result in the selection of a non-improving move.) The move is executed even when the move value is not positive, resulting in a deterioration of the current objective function value. The moved sector becomes tabu-active for a specified number of iterations (known as the *tabu tenure*), and therefore it cannot be selected for insertions during this time.

The Improvement Method terminates after *MaxInt* consecutive iterations without improvement. Before abandoning the improving phase, the *first*(N_2) procedure as described in Section 3 of Chapter 4 is applied to the best solution found during the current application of the Improvement Method. By applying this local search (without tabu restrictions), a local optimum with respect to the N_2 neighborhood is guaranteed as the output of the Improvement Method.

3.2.1   Computer Code

The SSImprove_solution function labeled Function 6-2 implements a short term memory tabu search. The search is initiated from the sol solution and the best solution found is stored in the best_sol during the search and copied onto sol when the tabu search finishes. The Iter variable controls the current iteration number, while the NonImproveIter keeps track of the number of consecutive iterations without improving best_sol.

The function starts with copying sol onto best_sol and value onto best_value. At each iteration the procedure probabilistically selects a sector j to perform a move. A random number a is uniformly generated in the interval [0, Sum] where Sum is the sum of all the weights in the matrix. The attractiveness of each sector is given by the sum of all the weights in its corresponding row and is stored in pb->row_sum. Each sector has an interval such that [0, row_sum(1)] is associated with sector 1,]row_sum(1), row_sum(1)+ row_sum(2)] is associated with sector 2, and so on. The selected sector is the one for which its associated interval contains the value of the random variable a falls is selected. If the sector is not tabu-active, the SSFirst_Insert function computes the first improving insertion for the selected sector j from its current position current_pos to the new position new_pos. If all the insertions of j deteriorate the objective function, SSFirst_Insert returns the best of them (i.e., that one with the smallest move value). The SSPerform_Move function executes the move and sector j becomes tabu active for tenure iterations. In this implementation we have fixed this tenure value to 10.

If the new solution improves the best solution found, the best solution is updated and the NonImproveIter variable is reset to 0. Otherwise the NonImproveIter is incremented by one. The procedure stops when NonImproveIter reaches MaxInt iterations. In this implementation, we have set MaxInt to 100. When the search ends, the best solution found best_sol is copied onto sol and its objective function value best_value is copied onto value. Finally, the SSLocalSearch function is called to perform a local search (without tabu restrictions) from the best solution found in order to reach a local optimum.

Function 6-2. SSImprove_solution — File: SSImprove3g.c

```
void SSImprove_solution(SS *pb,int sol[],int *value)
{
   int i,j,a,inc,current_pos,new_pos;
   int MaxInt=100,Iter,NonImproveIter=0;
   int *best_sol,*sol_inv,best_value,Sum=0;
   int *tabu,tenure=10;

   best_sol=SSInt_array(pb->nvar);
   tabu=SSInt_array(pb->nvar);
   sol_inv=SSInt_array(pb->nvar);
   for(i=1;i<=pb->nvar;i++)
     sol_inv[sol[i]]=i;

   /* Initialize Best Solution */
   best_value=*value;
   for(i=1;i<=pb->nvar;i++)
     best_sol[i]=sol[i];

   for(i=1;i<=pb->nvar;i++)
     Sum+=pb->row_sum[i];

   Iter=tenure+1;
   while(NonImproveIter <= MaxInt)
   {
```

```
Iter++;

/* Select a sector j according to row_sum */
a=SSGetrandom(0,Sum);
j=1;
while(a>pb->row_sum[j])
  a-=pb->row_sum[j++];

/* Check tabu status */
if(Iter-tabu[j]>tenure)
{
  /* Perform a move */
  current_pos=sol_inv[j];
  inc = SSFirst_Insert(pb,sol,current_pos,&new_pos);
  SSPerform_Move(sol,sol_inv,current_pos,new_pos);
  *value += inc;
  tabu[j]=Iter;

  if(*value>best_value)
  {
    NonImproveIter=0;
    best_value=*value;
    for(i=1;i<=pb->nvar;i++)
      best_sol[i]=sol[i];
  }
  else
    NonImproveIter++;
}
}

/* Copy best solution found */
*value=best_value;
for(i=1;i<=pb->nvar;i++)
  sol[i]=best_sol[i];

SSLocalSearch(pb,sol,value);

free(best_sol+1);free(tabu+1);
}
```

3.3   Reference Set

Another use of memory within scatter search is the one described in Section 1.2 of Chapter 1. Memory in this context is utilized to exclude reference solutions during the application of the Subset Generation Method. Recall that the Subset Generation Method keeps track of the subsets that have already been subjected to the Combination Method and therefore the method already employs memory as an integral part of its operation.

However, the memory mechanism described in Glover (1994) goes beyond excluding specific solution subsets to excluding all the subsets that include reference solutions that have a *tabu-active* status.

One immediate application of a tabu-like memory is connected with the 3-Tier Reference Set Update Method described in Section 1.2.3 of Chapter 5. Recall that this update partitions the reference set into three subsets: *RefSet₁*, *RefSet₂*, and *RefSet₃*. Generally speaking, the first subset has the best solutions found during the search, the second subset contains the most diverse solutions that were found during the search or were generated with the Diversification Generation Method and the third subset consists of "good generators" (i.e., solutions that have generated high quality solutions when combined with other solutions but that their quality is not sufficient to allow them to belong to the first subset). When a "good generator" is transferred from *RefSet₁* to *RefSet₃*, this solution could be given a tabu-active status for a short period of time (i.e., a small number of iterations). The rational for this tabu activation is that in many cases a reference solution that moves from *RefSet₁* to *RefSet₃* already has been combined with other solutions in the entire *RefSet*. Making the combination of this solution tabu for a specified tabu tenure, allows some changes in the *RefSet* and the future use of the recently transferred "good generator" more effective. This and other possible uses of memory in connection with the reference set deserve further consideration.

Chapter 7

CONNECTIONS WITH OTHER POPULATION-BASED APPROACHES

I believe there is something out there watching us. Unfortunately, it's the government.

Woody Allen

Throughout this book, we have established that scatter search (SS) belongs to the family of population-based metaheuristics. This family also includes the well-known evolutionary algorithms and the approach known as path relinking. In this chapter, we explore the connections between scatter search and genetic algorithms (GA), which is perhaps the best-known of the evolutionary algorithms. We also address the relationship between scatter search and path relinking.

Scatter search and genetic algorithms were both introduced in the seventies. While Holland (1975) introduced genetic algorithms and the notion of imitating nature and the "survival of the fittest" paradigm, Glover (1977) introduced scatter search as a heuristic for integer programming that expanded on the concept of surrogate constraints. Both methods are based on maintaining and evolving a set (population) of solutions throughout the search. Although the population-based approach makes SS and GA similar in nature, there are fundamental differences between these two methodologies. One main difference is that genetic algorithms were initially proposed as a mechanism to perform hyperplane sampling rather than optimization. Over the years, however, GAs have morphed into a methodology whose primary concern is the solution of optimization problems (Glover, 1994a).

In contrast, as detailed in Chapter 1, scatter search was conceived as an extension of a heuristic in the area of mathematical relaxation, which was designed for the solution of integer programming problems: surrogate constraint relaxation. The following three operations come from the area of mathematical relaxation and they are the core of most evolutionary algorithms including SS and GAs:

1. Building, maintaining and working with a population of elements (coded as vectors)
2. Creating new elements by combining existing elements
3. Determining which elements are retained based on a measure of quality

Two of the best-known mathematical relaxation procedures are Lagrangean relaxation (Everett, 1963) and surrogate constraint relaxation (Glover 1965). While Lagrangean approaches absorb "difficult" constraints into the objective function by creating linear combinations of them, surrogate constraint relaxation generate new constraints to replace those considered problematic. The generation of surrogate constraints also involves the combination of existing constraints using a vector of weights. In both cases, these relaxation procedures search for the best combination in an iterative manner. In Lagrangean relaxation, for example, the goal is to find the "best" combination, which, for a minimization problem, is the one that results in the smallest underestimation of the true objective function value. Since there is no systematic way of finding such weights (or so-called Lagrangean multipliers) in order to produce the smallest (possibly zero) duality gap, Lagrangean heuristics iteratively change the weights according to the degree of violation of the constraints that have been "brought up" to the objective function.

As intimated in Chapter 1, scatter search is more closely related to surrogate relaxation procedures, because not only surrogate relaxation includes the three operations outlined above but also has the goal of generating information from the application of these operations. In the case of surrogate relaxation, the goal is to generate information that cannot be extracted from the "parent constraints". Scatter search takes on the same approach, by generating information through combination of two or more solutions.

1.   GENETIC ALGORITHMS

The idea of applying the biological principle of natural evolution to artificial systems, introduced more than three decades ago, has seen impressive growth in the past few years. Usually grouped under the term evolutionary algorithms or evolutionary computation, we find the domains of genetic algorithms, evolution strategies, evolutionary programming, and genetic programming. Evolutionary algorithms have been successfully applied to numerous problems from different domains, including optimization, automatic programming, machine learning, economics, ecology, population genetics, studies of evolution and learning, and social systems.

A genetic algorithm is an iterative procedure that consists of a constant-size population of individuals, each one represented by a finite string of symbols, known as the genome, encoding a possible solution in a given problem space. This space, referred to as the search space, comprises all possible solutions to the problem at hand. Generally speaking, the genetic algorithm is applied to spaces that are too large to be exhaustively searched (such as those in combinatorial optimization). Solutions to a problem were originally encoded as binary strings due to certain computational advantages associated with such encoding. Also the theory about the behavior of algorithms was based on binary strings. Because in many instances it is impractical to represent solutions using binary strings, the solution representation has been extended in recent years to include character-based encoding, real-valued encoding, and tree representations.

The standard genetic algorithm proceeds as follows: an initial population of individuals is generated at random or heuristically. Every evolutionary step, known as a generation, the individuals in the current population are decoded and evaluated according to some predefined quality criterion, referred to as the fitness, or fitness function. To form a new population (the next generation), individuals are selected according to their fitness. Many selection procedures are currently in use, one of the simplest being Holland's original fitness-proportionate selection, where individuals are selected with a probability proportional to their relative fitness. This ensures that the expected number of times an individual is chosen is approximately proportional to its relative performance in the population. Thus, high-fitness ("good") individuals stand a better chance of "reproducing", while low-fitness ones are more likely to disappear.

The roulette-wheel selection is a common implementation of a proportional selection mechanism (Coello, et al. 2002). Figure 7-1 shows the roulette wheel associated with four individuals in a population with an average fitness of 100.

In this selection process, each individual in the population is assigned a portion of the wheel proportional to the ratio of its fitness and the population's average fitness. For example, individual A in Figure 7-1 is assigned half of the wheel because its fitness of 200 is twice the average fitness of 100. In a population of n individuals, an "average individual" (i.e., one with a fitness equal to the average fitness of the population) is assigned to $(100/n)\%$ of the wheel. For example, B is an average individual in the population of Figure 7-1.

Individual	Fitness
A	200
B	100
C	50
D	50
Mean	100

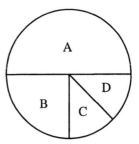

Figure 7-1. Roulette Wheel Selection

Tournament and ranking are two other popular selection techniques. Tournament selection consists of choosing q individuals from a population of n individuals and selecting the best according to the fitness value to survive into the next generation. Hence, n tournaments are necessary to build the population for the next generation. Binary tournaments, for which $q = 2$, are the most common implementation of this selection technique. Ranking ignores the fitness values and assigns selection probabilities based exclusively on rank.

Genetically inspired operators are used to introduce new individuals into the population, i.e., to generate new points in the search space. The best known of such operators are crossover and mutation. Crossover is performed, with a given probability p_c (the "crossover probability" or "crossover rate"), between two selected individuals, called parents, by exchanging parts of their genomes (i.e., encoding) to form two new individuals, called offspring; in its simplest form, sub-strings are exchanged after a randomly selected crossover point. This operator tends to enable the evolutionary process to move toward "promising" regions of the search space. Figure 7-2 depicts a one-point crossover operation.

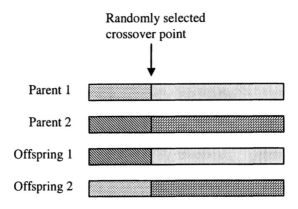

Figure 7-2. One-point crossover

The mutation operator is introduced to prevent premature convergence to local optima by randomly sampling new points in the search space. Mutation entails flipping bits at random, with some (small) probability p_m. Figure 7-3 shows a graphical representation of the mutation operator. Genetic algorithms are stochastic iterative processes that are not guaranteed to converge; the termination condition may be specified as some fixed, maximal number of generations or as the attainment of an acceptable fitness level for the best individual.

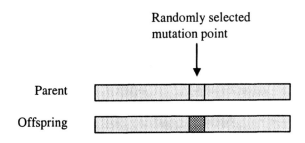

Figure 7-3. Mutation operator

Let us consider the following simple example to illustrate how genetic algorithms search. Suppose that the population consists of 4 individuals, which are binary-encoded strings (genomes) of length 8. The fitness value equals the number of ones in the bit string, with $p_c = 0.7$, and $p_m = 0.001$. More typical values of the population size and the genome length are in the range 50-1000. Also note that fitness computation in this case is extremely simple since neither complex decoding nor complex evaluation is necessary. The initial (randomly generated) population might look like the one shown in Table 7-1.

Table 7-1. Initial population

Label	Genome	Fitness
A	00000110	2
B	11101110	6
C	00100000	1
D	00110100	3

Using fitness-proportionate selection we must choose 4 individuals (two sets of parents), with probabilities proportional to their relative fitness values. In our example, suppose that the two parent pairs are (B, D) and (B, C). Note that A did not get selected as the procedure is probabilistic. Once a pair of parents is selected, the crossover operation is performed with probability p_c, resulting in two offspring. If the crossover operation is not performed (with probability $1-p_c$) then the offspring are exact copies of the parents. Suppose, in our example, that crossover takes place between parents B and D at the (randomly chosen) first bit position, forming offspring E=10110100 and F=01101110. Also suppose that the crossover operation is not performed between parents B and C, so the offspring are exact copies of B and C. Next, each offspring is subject to mutation with probability p_m per bit. For example, suppose offspring E is mutated at the sixth position to form E'=10110000, offspring B is mutated at the first bit position to form B'=01101110, and offspring F and C are not mutated at all. The next generation population, created by the above operators of selection, crossover, and mutation is shown in Table 7-2.

Table 7-2. Population after one generation

Label	Genome	Fitness
E'	10110000	3
F	01101110	5
C	00100000	1
B'	01101110	5

Note that in the new population in Table 7-2, although the best individual with fitness 6 has been lost, the average fitness has increased. Iterating this procedure, the genetic algorithm would eventually find, for this small example, a perfect string, i.e., with maximal fitness value of 8. More sophisticated implementations of GAs include the use of local search and several crossover operators that are chosen probabilistically to be applied to each pair of selected parents.

1.1   SS and GA Comparison

There are some obvious differences between genetic algorithms and scatter search:

1. The population in genetic algorithms is about one order of magnitude larger than the reference set in scatter search. A typical GA population size is 100, while a typical SS reference set size is 10.
2. A probabilistic procedure is used to select parents to apply crossover and mutation operators in GAs while the Combination Method is applied to a predetermined list of subsets of reference solutions in scatter search.
3. The evolution of a GA population follows the "survival of the fittest" philosophy, which is implemented using probabilistic rules. In scatter search, changes in the reference set are controlled by the deterministic rules in the Reference Set Update Method.

4. The use of local search procedures is an integral part of scatter search, while it was added to GAs in order to create hybrid implementations that would yield improved outcomes.
5. The Subset Generation Method considers combinations of more than two solutions. GAs are typically restricted to combining two solutions.
6. Full randomization is the typical mechanism used in GAs to build the initial population. Diversity in scatter search is achieved with the application of the Diversification Generation Method, which is designed to balance diversity with solution quality.

Although GAs and SS have contrasting views about searching a solution space, it is possible to create a hybrid approach without entirely compromising the scatter search framework. Specifically, if we view the crossover and mutation operators as an instance of a Combination Method, it is then straightforward to design a scatter search procedure that employs genetic operators for combination purposes.

Martí, Laguna and Campos (2002) compare the performance of scatter search and genetic algorithms employing four classes of problems whose solutions can be represented with a permutation. The SS and GA implementations are based on a model that treats the objective function evaluation as a black box, making the search procedures context-independent. This means that neither implementation takes advantage of the structural details of the tests problems. The comparisons are based on experimental testing with four well-known problems: linear ordering, traveling salesperson, matrix bandwidth reduction and a job-sequencing problem. The only information that both the SS solver and the GA solver have with respect to these problems is the nature of the objective function evaluation with regard to the "absolute" or "relative" positioning of the elements in the permutation. In other words, the problems are classified into two classes:

- *A-permutation problems*—for which absolute positioning of the elements is more important (e.g., linear ordering problem)

- *R-permutation problems*—for which relative positioning of the elements is more important (e.g., traveling salesperson problem)

Not all problems can be fully characterized as "absolute" or "relative"; however, this does not render the proposed implementations useless. The key to this computational testing and comparison is that both methods search using the strategies to combine solutions and generate new ones. They also use the same procedure to improve new trial solutions.

The scatter search implementation follows the structure outlined in Figure 2-1 of Chapter 2. The Diversification Generation Method is the systematic procedure described in Glover (1998). The distance between two permutations $p = (p_1, p_2, ..., p_n)$ and $q = (q_1, q_2, ..., q_n)$ depends on the type of problem being solved. For A-permutation problems, the distance is given by:

$$d(p,q) = \sum_{i=1}^{n} |p_i - q_i|.$$

The distance for R-permutation problems is defined as:

$d(p,q) =$ number of times p_{i+1} does not immediately follow p_i in q, for
               $i = 1, ..., n\text{-}1$

The Subset Generation Method is limited to the generation of all reference solution pairs with at least one new solution. The Reference Set Update Method builds the initial reference set with $b/2$ high quality solutions and $b/2$ diverse solutions, where $b = |RefSet|$. The updating during the search is based on quality only. The Combination Method is applied to all pairs of solutions in the current *RefSet*.

The genetic algorithm is also a standard implementation. It follows the scheme of Michalewicz (1996) and is summarized in Figure 7-4.

---

1. *Generate solutions* — Build P by randomly generating *PopSize* permutations.
2. *Improve solutions* — Apply the local search method to improve solutions generated in Step 1.

while (objective function evaluations < *MaxEval*) **do** {

    3. *Evaluate solutions* — Evaluate the solutions in P and update the best solution found if necessary.

    4. *Survival of the fittest* — Calculate the probability of surviving based on solution quality. Evolve P by choosing *PopSize* solutions according to their probability of surviving.

    5. *Combine solutions* — Select a fraction p_c of the solutions in P to be combined. Selection is at random with the same probability for each element of P. The selected elements are randomly paired for combination, with each pair generating two offspring that replace their parents in P.

    6. *Mutate solutions* — A fraction p_m of the solutions in P is selected for mutation. The mutated solution replaces the original in P.

}

Figure 7-4. GA outline

The outline in Figure 7-4 indicates that the GA implementation maintains a population *P* of solutions and the search is performed until *MaxEval* objective function evaluations. The same stopping criterion is used in the scatter search implementation. To provide for a fair comparison, both procedures use the same Combination and Improvement Methods, as described in the following subsections.

1.1.1 Improvement Method

Insertions are used as the primary mechanism to move from one solution to another in the Improvement Method common to both the GA and SS implementations. Let $MOVE(p_j, i)$ consist of deleting p_j from its current position j in p to be inserted in position i (i.e., between the elements p_{i-1} and p_i that are currently in positions i-1 and i). This operation results in the ordering p' as follows:

$$p' = \begin{cases} \left(p_1, \ldots, p_{i-1}, p_j, p_i, \ldots, p_{j-1}, p_{j+1}, \ldots, p_n\right) & \text{for } i < j \\ \left(p_1, \ldots, p_{j-1}, p_{j+1}, \ldots, p_i, p_j, p_{i+1}, \ldots, p_n\right) & \text{for } i > j \end{cases}$$

Since the local search method is context independent, the only available mechanism for computing the move value is submitting p' for evaluation and comparing its value with the value of p. In order to reduce the computational effort associated with evaluating moves for possible selection and to increase the search efficiency, $INSERT(p_j)$ is defined as the set of promising insert positions for p_j. The local search considers inserting p_j only in those positions in $INSERT(p_j)$. Then, the neighborhood N of the current solution is given as:

$$N = \{p' : MOVE(p_j, i), \text{ for } j = 1, \ldots, n \text{ and } i \in INSERT(p_j)\}$$

The neighborhood N is partitioned into n sub-neighborhoods N_j associated with each element p_j as:

$$N_j = \{p' : MOVE(p_j, i), \ i \in INSERT(p_j)\}$$

The set $INSERT(p_j)$ depends on whether the problem is an A-permutation or a R-permutation problem. In A-permutation problems the procedure accumulates in $FreqIns(i,j)$ the number of times that element i has been inserted in position j improving the current solution. Then, given an element i, it computes $m(i)$ as the position j where the value of $FreqIns(i,j)$ is maximal. The search considers that $m(i)$ and the positions around it are desirable positions for inserting element i. This information is used to assign $INSERT(p_j)$ the following values:

$$INSERT(p_j) = [\ m(p_j) - RANGE,\ m(p_j) + RANGE\ ]$$

The value of *RANGE* is an additional search parameter. In R-permutation problems the procedure accumulates in $FreqIns(i,j)$ the number of times that element i has been inserted in the position immediately preceding element j. Then it computes $m(i)$ as the element j with maximal $FreqIns(i,j)$ value. Let $pp(e)$ be the previous position of element e in the current solution. Then the search considers that $pp(m(i))$ is a desirable position for inserting element i. In this case, $INSERT(p_j)$ is assigned the following values:

$$INSERT(p_j) = \{\ pp(e)\ /\ FreqIns(p_j,e) \geq \alpha\ m(p_j)\ \}$$

where the value of α is dynamically adjusted to obtain a set with $2*RANGE$ elements. The implementation is such that it avoids ordering all the elements in $FreqIns(i,j)$, so it is able to construct the set $INSERT(p_j)$ with a low computational effort.

The rule for selecting an element for insertion is based on frequency information. Specifically, the number of times that element j has been moved resulting on an improved solution is accumulated in $freq(j)$. The probability of selecting element j is proportional to its frequency value $freq(j)$.

Starting from a new trial solution the procedure chooses the best insertion associated with a given element. At each iteration, an element p_j in the current solution p is probabilistically selected according to its $freq(j)$ value. The solution p' with the lowest value in N_j is selected. The local search procedure execute only improving moves. An improving move is one for which the objective function value of p' is better (strictly smaller for minimization problems or strictly larger for maximization problems) than the objective function value of p. The local search terminates when no improving move is found after *NTrials* elements are consecutively selected and the exploration of their neighborhood fails to find an improving move.

1.1.2   Combination Method

The Combination Method is as important to scatter search implementations as the crossover and mutation operators are to genetic algorithms. Combination methods are typically adapted to the problem context in the same way that crossover operators have been progressively specialized to specific settings. For instance, the combination methods described in the tutorial chapters were adapted to each particular problem context. In order to design a context-independent combination methodology that performs well across a wide collection of different permutation problems, the Combination Method consists of ten combination strategies from which one is probabilistically selected according to its performance in previous iterations during the search.

In the scatter search implementation, solutions in the *RefSet* are ordered according to their objective function value. So, the best solution is the first one in *RefSet* and the worst is the last one. When a solution obtained with combination strategy i (referred to as cs_i) qualifies to be the jth member of the current *RefSet*, the quantity b-j+1 is added to $score(cs_i)$. Therefore, combination strategies that generate good solutions accumulate higher scores and in turn increase their probability of being selected. To avoid initial biases, this mechanism is activated after the first *InitIter* combinations, and before this point the selection of the combination strategy is made completely at random.

In the GA implementation, when a solution obtained with combination strategy cs_i is better than its parent solutions, $score(cs_i)$ is increased by one. If the combination strategy is a mutation operator then the score is increased by one when the mutated solution is better than the original solution. Experimentation with this strategy in the GA implementation showed that there was no significant difference between using the scheme from the beginning of the search and waiting *InitIter* combinations to activate it. Therefore, the probabilistic selection procedure is activated from the beginning of the GA search. A description of the ten combination strategies follows, where the solutions being combined are referred to as "reference solutions" and to the resulting solutions as the "new trial solutions" (although in the GA literature reference solutions are known as "parents" and the new trial solutions as "offspring").

Combination Strategy 1

This is an implementation of a classical GA crossover operator. The method randomly selects a position k to be the crossing point from the range $[1, n/2]$. The first k elements are copied from one reference solution while the remaining elements are randomly selected from both reference solutions. For each position i ($i = k+1, \ldots, n$) the method randomly selects one reference solution and copies the first element that is still not included in the new trial solution.

Combination Strategy 2
This strategy is a special case of 1, where the crossing point k is always fixed to one.

Combination Strategy 3
This is an implementation of what is known in the GA literature as the partially matched crossover. The strategy is to randomly choose two crossover points in one reference solution and copy the partial permutation between them into the new trial solution. The remaining elements are copied from the other reference solution preserving their relative ordering.

Combination Strategy 4
This method is a case of what is referred to in the GA literature as a mutation operator. The strategy consists of selecting two random points in a chosen reference solution and inverting the partial permutation between them. The inverted partial permutation is copied into the new trial solution. The remaining elements are directly copied from the reference solution preserving their relative order.

Combination Strategy 5
This combination strategy also operates on a single reference solution. It consists of scrambling a randomly selected portion of the reference solution. The remaining elements are directly copied from the reference solution into the new trial solution.

Combination Strategy 6
This is a special case of combination strategy 5 where the partial permutation to be scrambled starts in position 1 and the length is randomly selected in the range $[2, n/2]$.

Combination Strategy 7
The strategy is the Combination Method described in Section 5 of Chapter 4. The method scans (from left to right) both reference solutions, and uses the rule that each reference solution votes for its first element that is still not included in the new trial solution (referred to as the "incipient element"). The voting determines the next element to enter the first still unassigned position of the trial solution. This is a min-max rule in the sense that if any element of the reference solution is chosen other than the incipient element, then it would increase the deviation between the reference and the trial solutions. Similarly, if the incipient element were placed later in the trial solution than its next available position, this deviation would also increase. So the rule attempts to minimize the maximum deviation of the

trial solution from the reference solution under consideration, subject to the fact that other reference solution is also competing to contribute. A bias factor that gives more weight to the vote of the reference solution with higher quality is also implemented for tie breaking. This rule is used when more than one element receives the same votes. Then the element with highest weighted vote is selected, where the weight of a vote is directly proportional to the objective function value of the corresponding reference solution.

Combination Strategy 8
In this strategy the two reference solutions vote for their incipient element to be included in the first still unassigned position of the trial solution. If both solutions vote for the same element, the element is assigned. If the reference solutions vote for different elements but these elements occupy the same position in both reference permutations, then the element from the permutation with the better objective function is chosen. Finally, if the elements are different and occupy different positions, then the one in the lower position is selected.

Combination Strategy 9
Given two reference solutions p and q, this method probabilistically selects the first element from one of these solutions. The selection is biased by the objective function value corresponding to p and q. Let e be the last element added to the new trial solution. Then, p votes for the first unassigned element that is positioned after e in the permutation p. Similarly, q votes for the first unassigned element that is positioned after e in q. If both reference solutions vote for the same element, the element is assigned to the next position in the new trial solution. If the elements are different then the selection is probabilistically biased by the objective function values of p and q.

Combination Strategy 10
This is a deterministic version of combination strategy 9. The first element is chosen from the reference solution with the better objective function value. Then the reference solutions vote for the first unassigned successor of the last element assigned to the new trial solution. If both solutions vote for the same element, then the element is assigned to the new trial solution. Otherwise, the "winner" element is determined with a score, which is updated separately for each reference solution in the combination. The score values attempt to keep the proportion of times that a reference solution "wins" close to its relative importance, where the importance is measured by the value of the objective function. The scores are calculated to

minimize the deviation between the "winning rate" and the "relative importance". For example, if two reference solutions p and q have objective function values of $value(p) = 40$ and $value(q) = 60$, then p should contribute with 40% of the elements in the new trial solution and q with the remaining 60% in a maximization problem. The scores are updated so after all the assignments are made, the relative contribution from each reference solution approximates the target proportion. More details about this combination strategy can be found in Glover (1994b).

1.1.3   Computational Testing

The performance comparison between SS and GA is based on experiments with four classes of combinatorial optimization problems:

- the bandwidth reduction problem
- the linear ordering problem
- the traveling salesman problem
- a single machine sequencing problem

These problems are a good test set because they are well known, they are different in nature and problem instances with known optimal or high-quality solutions are readily available. Existing methods to solve these problems range from construction heuristics and metaheuristics to exact procedures.

The *bandwidth reduction problem* (BRP) has the goal of finding a configuration of a matrix that minimizes its bandwidth. The bandwidth of a matrix $A = \left\{ a_{ij} \right\}$ is defined as the maximum absolute difference between i and j for which $a_{ij} \neq 0$. The BRP consists of finding a permutation of the rows and columns that keeps the nonzero elements in a band that is as close as possible to the main diagonal of the matrix; the objective is to minimize the bandwidth. This NP-hard problem can also be formulated as a labeling of vertices on a graph, where edges are the nonzero elements of the corresponding symmetrical matrix. Metaheuristics proposed for this problem include a simulated annealing implementation by Dueck and Jeffs (1995) and a tabu search approach by Martí, et al. (2001).

The Linear Ordering Problem (LOP) was described in Chapter 4. The Traveling Salesman Problem (TSP) consists of finding a tour (cyclic permutation) visiting a set of cities that minimizes the total travel distance. An incredibly large amount of research has been devoted to this problem, which would be absurd and impractical to summarize here. However, valuable references about the TSP include Lawler, et al (1985), Reinelt (1994) and Gutin and Punnen (2002).

Finally, the fourth problem is a single machine-sequencing problem (SMSP) with delay penalties and setup costs. At time zero, *n* jobs arrive at a continuously available machine. Each job requires a specified number of time units on the machine and a penalty (job dependent) is charged for each unit that job commencement is delayed after time zero. In addition, there is a setup cost s_{ij} charged for scheduling job *j* immediately after job *i*. The objective is to find the schedule that minimizes the sum of the delay and setup costs for all jobs. Note that if delay penalties are ignored, the problem becomes an asymmetric traveling salesman problem. Barnes and Vanston (1981) reported results of three branch and bound algorithms applied to instances with up 20 jobs and Laguna, Barnes and Glover (1993) developed a TS method that was tested in a set of instances whose size ranged between 20 and 35 jobs.

We employed the following problem instances for computational testing. The instances are included in the Instances directory of the Chapter7 folder in the accompanying disc:

- 37 BRP instances from the Harwell-Boeing Sparse Matrix Collection. This collection consists of a set of standard test matrices associated with linear systems, least squares, and eigenvalue calculations from a wide variety of scientific and engineering disciplines. The size of these instances ranges between 54 and 685 rows with an average of 242.9 rows (and columns).
- 49 LOP instances from the public-domain library LOLIB (1997). These instances consist of input-output tables from economic sectors in the European community and their size ranges between 44 and 60 rows with an average of 48.5 rows (and columns).
- 31 TSP instances from the public-domain library TSPLIB (1995). These instances range in size between 51 and 575 cities with an average of 159.6 cities.
- 40 SMS instances from Laguna, Barnes and Glover (1993). The best solutions available for these instances are not proven optimal, however they are the best upper bounds ever found. These instances range in size between 20 and 35 jobs with an average of 26.0 jobs.

In order to apply the different strategies described above, the BRP and SMSP were classified as A-permutation problems and the LOP and TSP as R-permutation problems. Note that the objective function in the SMSP is influenced by both the absolute position of the jobs (due to the delay penalties) and the relative position of the jobs (due to the setup costs). The classification is based on the knowledge that the delay penalties in the tested instances are relatively larger when compare to the setup cost. In cases when this knowledge is not available, it would be recommended to run the procedure twice on a sample set of problems in order to establish the relative importance of the positioning of elements in the permutation.

The solution procedures were implemented in C++ and compiled with Microsoft Visual C++ 6.0, optimized for maximum speed. All experiments were performed on a Pentium III at 800 MHz. The scatter search parameters *PopSize*, *b* and *InitIter* were set to 100, 10 and 50 respectively, as recommended in Campos et al. (1999). The parameters associated with the local search procedure were set after some preliminary experimentation to *RANGE* = 3 and *Ntrials* = 25. The GA parameters were set to *PopSize* = 100, p_c = 0.25 and p_m = 0.01, as recommended in Michalewicz (1996). Both procedures used the stopping criterion of 1 million objective function evaluations.

In this set of experiments the SS implementation is compared to two versions of the GA procedure: 1) without local search (GA) and 2) with local search (GALS). SS always uses local search by design. The two GA versions complement each other in that one uses part of its objective function evaluation "budget" to perform local searches and the other uses its entire budget to apply the combination operators and evolve the population of solutions. In the first set of experiments the use of combination strategies is restricted in such a way that both GA versions use combination strategies 1-6 and SS uses combination strategies 7-10. Recall that combination strategies 1-3 are crossover operators and combination strategies 4-7 are mutation operators. Combination strategies 7-10 are more strategic in nature and one is completely deterministic.

Table 7-3 shows the average percent deviation from the best-known solution to each problem instance. The procedures were executed once with a fixed random seed and the average is over all instances. It should be mentioned that in the case of LOP and TSP the best solutions considered are the optimal solutions as given in the public libraries. In the case of the BRP the best solutions are from Martí et al (2001) and the best solutions for the SMSP instances are due to Laguna et al (1993).

Table 7-3. Percent deviation from best solutions

Method	BRP	LOP	SMSP	TSP
GA	59.124%	6.081%	4.895%	95.753%
GALS	55.500%	0.005%	0.832%	143.881%
SS	58.642%	0.000%	0.291%	43.275%

Table 7-3 shows that SS performs better on average than GA in all problem types and that it is inferior to GALS when tackling BRP instances. The three methods are able to obtain high quality results for the LOP and the SMSP instances. The results for the TSP are less desirable, but it should be noted that these problems are on average larger than the LOP and SMSP instances. Table 7-4 presents the average percent improvement of SS over the two GA versions. The largest improvement occurs when solving TSP instances and the negative improvement associated with GALS shows the better average performance of this method when compared to SS.

Table 7-4. Percent improvement of SS over GA and GALS

Method	BRP	LOP	SMS	TSP
GA	0.3%	6.5%	4.4%	26.8%
GALS	-2.0%	0.0%	0.5%	41.3%

Table 7-4 shows that including local search within the GA framework yields better results than without it, except in the TSP case. For the TSP, the performance of the GA implementation deteriorates when coupled with the local search procedure. The SS vs. GALS in the case of BRP instances deserves a closer examination, because it is the only case in which either one of the GA implementations outperforms SS. Figure 7-5 shows the trajectory of the average percent deviation from the best-known solution as the SS and GALS searches progress.

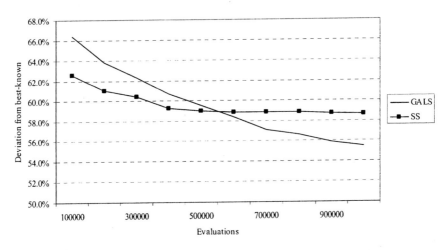

Figure 7-5. Percent deviation trajectory for SS and GALS solving BRP instances

The percent deviation value trajectory in Figure 7-5 shows that while SS results are better than or at least as good as GALS results before 600 thousand evaluations, GALS keeps improving while SS stagnates. Our explanation for this behavior is as follows. In the BRP, the change in the objective function value from one solution to another does not represent a meaningful guidance for the search. In particular, the objective is a min-max function that in many cases results in the same evaluation for all the solutions in the neighborhood of a given solution. Since SS relies on strategic choices, it is unable to obtain information from the evaluation of this "flat landscape" function to direct the search. GALS, heavily relying on randomization, is able to more effectively probe the solution space presented by BRP instances.

Figure 7-6 depicts the trajectory of the percent deviation from the best-known solution to the SMSP as the search progress. The average values of the three methods under consideration are shown.

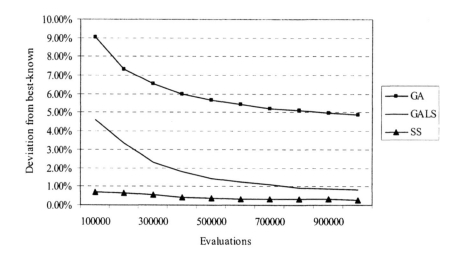

Figure 7-6. Percent deviation trajectory when solving SMS instances

Figure 7-6 shows that the SS procedure is able to obtain high quality solutions from the very beginning of the search. Specifically, after 100,000 objective function evaluations, the percent deviation from the best-known solutions for the SS method is 0.7 while after 1 million evaluations the GA and the GALS methods are still at 4.8 and 0.8, respectively. The strategic choices of SS in this case make a big performance difference when compared to the two GA variants.

In the second experiment, SS, GA and GALS are compared when allowed to use all the combination strategies described in Section 1.1.2 of this chapter. The results of these experiments are summarized in Tables 7-5 and 7-6. These tables are equivalent to Tables 7-3 and 7-4 in that they shows the percent deviation from the best known solution values and the percent improvement of SS over GA and GALS, respectively.

Table 7-5. Percent deviation from best

Method	BRP	LOP	SMSP	TSP
GA	58.244%	6.722%	4.940%	101.689%
GALS	55.102%	0.004%	0.268%	133.792%
SS	52.587%	0.000%	0.207%	54.321%

Table 7-6. Percent improvement of SS over GA and GALS

Method	BRP	LOP	SMSP	TSP
GA	3.57%	7.21%	4.51%	23.49%
GALS	1.62%	0.00%	0.06%	33.99%

A direct comparison of Tables 7-3 and 7-5 shows the advantage of using all combination strategies with both the GA and SS implementations. Tables 7-5 and 7-6 indicate that when all combination strategies are used, SS has a an average performance that is superior to GA and GALS. The performance is only marginally better in the case of the LOP and SMSP, but it continues to be significantly better in the case of TSP. When using all operators, including those that incorporate a fair amount of randomization, SS is able to outperform GALS in the BRP instances, as shown in Figure 7-7.

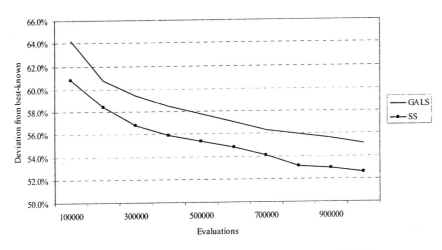

Figure 7-7. Percent deviation trajectory for SS and GALS solving BRP instances

Figure 7-8 shows the trajectory of the percent deviation from the best-known solutions to the SMSP as the search progress. The average values shown in Figure 7-8 correspond to the second experiment, where all combination methods are made available to SS, GA and GALS.

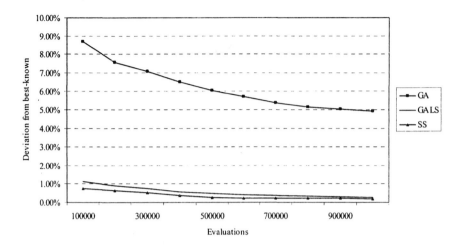

Figure 7-8. Percent deviation trajectory when solving SMSP instances

A comparison of Figures 7-6 and 7-8 reveals that while SS shows a minor improvement when solving SMSP instances using all combination strategies, the improvement associated with GALS is significant. GALS achieves a deviation of 1.14% at the 100,000 evaluation mark when applying all 10 combination strategies, which compares quite favorably with a deviation of 4.8% at the same stage of the search when using a fraction of the available combination strategies. Mixing combination strategies with random elements and those based on systematic mechanisms seems to benefit procedures based on the GA framework and those based on the SS methodology.

2. PATH RELINKING

Path relinking (PR) was originally suggested as an approach to integrate intensification and diversification strategies in the context of tabu search (Glover, 1994b; Glover and Laguna, 1997). This approach generates new solutions by exploring trajectories that connect high-quality solutions, by starting from one of these solutions, called an *initiating solution*, and generating a path in the neighbourhood space that leads toward the other solutions, called *guiding solutions*. This is accomplished by selecting moves that introduce attributes contained in the guiding solutions.

Path relinking can be considered an extension of the Combination Method of scatter search. Instead of directly producing a new solution when combining two or more original solutions, PR generates paths between and beyond the selected solutions in the neighborhood space. The character of such paths is easily specified by reference to solution attributes that are added, dropped or otherwise modified by the moves executed. Examples of such attributes include edges and nodes of a graph, sequence positions in a schedule, vectors contained in linear programming basic solutions, and values of variables and functions of variables.

The approach may be viewed as an extreme (highly focused) instance of a strategy that seeks to incorporate attributes of high quality solutions, by creating inducements to favor these attributes in the moves selected. However, instead of using an inducement that merely encourages the inclusion of such attributes, the path relinking approach subordinates other considerations to the goal of choosing moves that introduce the attributes of the guiding solutions, in order to create a "good attribute composition" in the current solution. The composition at each step is determined by choosing the best move, using customary choice criteria, from a restricted set — the set of those moves currently available that incorporate a maximum number (or a maximum weighted value) of the attributes of the guiding solutions.

The approach is called path relinking either by virtue of generating a new path between solutions previously linked by a series of moves executed during a search, or by generating a path between solutions previously linked to other solutions but not to each other. Figure 7-9 shows two hypothetical paths (i.e., a sequence of moves) that link solution A to solution B, to illustrate relinking of the first type. The solid line indicates an original path produced by the "normal" operation of a procedure that produced a series of moves leading from A to B, while the dashed line depicts the relinking path. The paths are different because the move selection during the normal operation does not "know" where solution B lies until it is finally reached, but simply follows a trajectory whose intermediate steps are determined by some form of evaluation function. For example, a commonly used approach is to select a move that minimizes (or maximizes) the objective function value in the local sense. During path relinking, however, the main goal is to incorporate attributes of the guiding solution (or solutions) while at the same time recording the objective function values.

The effort to represent the process in a simple diagram such as the one in Figure 7-9 may create some misleading impressions. First, the original (solid line) path, which is shown to be "greedy" relative to the objective function, is likely to be significantly more circuitous along dimensions we are not able to show, and by the same token to involve significantly more steps (intervening solutions) not portrayed in Figure 7-9. Second, because the relinked path is not governed so strongly by local attraction, but instead is influenced by the criterion of incorporating attributes of the guiding solution, it opens the possibility of reaching improved solutions that would not be found by a "locally myopic" search. Figure 7-9 shows one such solution (the darkened node) reached by the dotted path. Beyond this, however, the relinked path may encounter solutions that may not be better than the initiating or guiding solution, but that provide fertile "points of access" for reaching other, somewhat better, solutions. For this reason it is valuable to examine neighboring solutions along a relinked path, and keep track of those of high quality which may provide a starting point for launching additional searches.

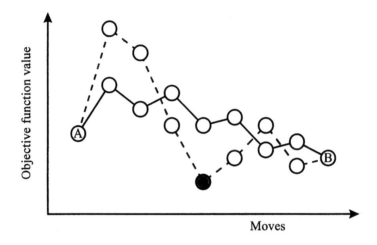

Figure 7-9. Path relinking illustration

To generate the desired paths, it is only necessary to select moves that perform the following role: upon starting from an *initiating solution,* the moves must progressively introduce attributes contributed by a *guiding solution* (or reduce the distance between attributes of the initiating and guiding solutions). The roles of the initiating and guiding solutions are interchangeable; each solution can also be induced to move simultaneously toward the other as a way of generating combinations. First consider the creation of paths that join two selected solutions x' and x'', restricting attention to the part of the path that lies 'between' the solutions, producing a solution sequence $x' = x(1), x(2), ..., x(r) = x''$. To reduce the number of options to be considered, the solution $x(i + 1)$ may be created from $x(i)$ at each step by choosing a move that minimizes the number of moves remaining to reach x''.

Instead of building a reference set (*RefSet*) as in scatter search, path relinking usually starts from a given set of elite solutions obtained during a search process. To simplify the terminology, we will also let *RefSet* refer to this set of b solutions that have been selected during the application of the embedded search method. This method can be Tabu Search, as in Laguna, Martí and Campos (1999), GRASP, as in Laguna and Martí (1999), or simply a Diversification Generation Method coupled with an Improvement Method as proposed in scatter search. From this point of view, SS and PR can be considered population-based methods that operate on a set of reference solutions and basically differ in the way in which the reference set is constructed, maintained, updated and improved.

In basic scatter search designs, all pairs of solutions in the *RefSet* are subjected to the Combination Method. Similarly, in a basic version of PR all pairs in the *RefSet* are considered to perform a relinking phase. For each pair (x', x'') two paths are initiated; one from x' to x'' and the other from x'' to x'.

Several studies have experimentally found that it is convenient to add a local search exploration from some of the generated solutions within the relinking path, as proposed in Glover (1994b), in order to produce improved outcomes. We refer the reader to Laguna and Martí (1999), Piñana et al. (2001) or Laguna, Martí and Campos (1999) for some examples. Note that two consecutive solutions after a relinking step are very similar and differ only in the attribute that was just introduced. Therefore, it is generally not efficient to apply an Improvement Method at every step of the relinking process. We introduce the parameter *NumImp* to control its application. In particular, the Improvement Method is applied every *NumImp* steps of the relinking process. An alternative suggested in Glover (1994b) is to keep track of a few "best solutions" generated during the path trace, or of a few best neighbors of the solutions generated, and then return to these preferred candidate solutions to initiate the improvement process.

Figure 7-10 shows a simple PR procedure for a minimization problem. It starts with the creation of an initial set of *b* elite solutions (*RefSet*). As in scatter search, the solutions in *RefSet* are ordered according to quality, and the search is initiated by assigning the value of TRUE to *NewSolutions*. In step 3, *NewSubsets* is constructed with all the pairs of solutions in *RefSet*, and *NewSolutions* is switched to FALSE. Also in this step, *Pool* is initialized to empty. The pairs in *NewSubsets* are selected one at a time in lexicographical order and the Relinking Method is applied to generate two paths of solutions in steps 5 and 7. The solutions generated in these steps are added to *Pool*. The Improvement Method is applied every *NumImp* steps of the relinking process in each path (steps 6 and 8). Solutions found during the application of the Improvement Method are also added to *Pool*. Each solution in Pool is examined to see whether it improves upon the worst solution currently in *RefSet*. If so, the new solution replaces the worst and *RefSet* is reordered in step 9. The *NewSolutions* flag is switched to TRUE in step 10 and the pair (x', x'') that was just combined is deleted from *NewSubsets* in step 11.

1.   Obtain a *RefSet* of *b* elite solutions.
2.   Evaluate the solutions in *RefSet* and order them according to their objective function
     value such that x^1 is the best solution and x^b the worst. Make *NewSolutions* = TRUE.
     while (*NewSolutions*) **do**
      3.   Generate *NewSubsets*, which consists of all pairs of solutions in *RefSet* that
          include at least one new solution. Make *NewSolutions* = FALSE and
          Pool = $\varnothing$.
    while (*NewSubsets* $\neq \varnothing$) **do**
        4.   Select a next pair (x', x'') in *NewSubSets*.
        5.   Apply the Relinking Method to produce the sequence $x' = x'(1), x'(2), \ldots,$
          $x'(r) = x''$ and add solutions to *Pool*.

    for $i = 1$ **to** $i < \dfrac{r}{NumImp}$ **do**
        6.   Apply the Improvement Method to $x'(i*NumImp)$ and add solutions
          to *Pool*.
    end for
        7.   Apply the Relinking Method to produce the sequence $x'' = x''(1), x''(2),$
          $\ldots, x''(s) = x'$ and add solutions to *Pool*.

    for $i = 1$ **to** $i < \dfrac{s}{NumImp}$ **do**
        8.   Apply the Improvement Method to $x''(i*NumImp)$ and add solutions
          to *Pool*.
    end for
    for (each solution $x \in$ *Pool*)
        if ($x \notin$ *RefSet* and $f(x) < f(x^b)$) **then**
          9.   Make $x^b = x$ and reorder *RefSet*
          10. Make *NewSolutions* = TRUE
        end if
    end for
        11. Delete (x', x'') from *NewSubsets*
    end while
  end while

Figure 7-10. Outline of a simple path relinking procedure

As in the basic scatter search designs, the updating of the reference set in Figure 7-10 is based on improving the quality of the worst solution and the search terminates when no new solutions are admitted to *RefSet*. Similarly, the Subset Generation Method is also very simple and consists of generating all pairs of solutions in *RefSet* that contain at least one new solution. We examine strategies to overcome the limitations of this basic design in the next subsections.

It is possible, as in applying scatter search, that x' and x'' were previously joined by a search trajectory produced by a heuristic method (or by a metaheuristic such as tabu search). In this event, the new trajectory created by path relinking is likely to be somewhat different than the one initially established, as illustrated in Figure 7-9.

It may also be that x' and x'' were not previously joined by a search path at all, but were generated on different search paths, which may have been produced either by a heuristic or by a previous relinking process. Such a situation is depicted in Figure 7-11. In this case, the path between x' and x'' performs a relinking function by changing the connections that generated x' and x'' originally.

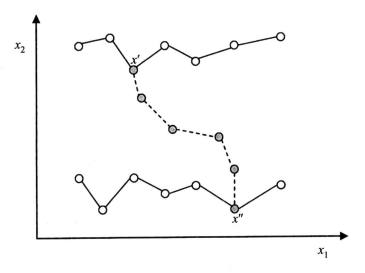

Figure 7-11. Previously generated paths shown by heavy lines and relinked path shown by dotted line

To choose among the different paths that may be possible in going from x' to x'', let $f(x)$ denote an objective function which is to be minimized. Selecting unattractive moves relative to $f(x)$, from the moves that are candidates to generate the path at each step, will tend to produce a final series of strongly improving moves to complete the path. Correspondingly, selecting attractive moves at each step will tend to produce lower quality moves at the end. (The last move, however, will be improving, or leave $f(x)$ unchanged, if x'' is selected to be a local optimum.) Thus, choosing best, worst or average moves, provides options that produce contrasting effects in generating the indicated sequence. An aspiration criterion may be used as in tabu search to override choices in the last two cases if a sufficiently attractive solution is available. (In general, it appears reasonable to select best moves at each step, and then to allow the option of reinitiating the process in the opposite direction by interchanging x' and x''.)

The choice of one or more solutions $x(i)$ to become reference points for launching a new search phase will preferably be made to depend not only on $f(x(i))$ but also on the $f(x)$ values of those solutions x that can be reached by a move from $x(i)$. The process can additionally be varied to allow solutions to be evaluated other than those that yield $x(i + 1)$ closer to x''. Aspiration criteria again are relevant for deciding whether such solutions qualify as candidates for selection.

To elaborate the process, let $x^*(i)$ denote a neighbor of $x(i)$ that yields a minimum $f(x)$ value during an evaluation step, excluding $x^*(i) = x(i + 1)$. If the choice rules do not automatically eliminate the possibility $x^*(i) = x(h)$ for $h < i$, then a simple tabu restriction can be used to do this (e.g., see Glover and Laguna, 1997). Then the method selects a solution $x^*(i)$ that yields a minimum value for $f(x^*(i))$ as a new point to launch the search. If only a limited set of neighbors of $x(i)$ are examined to identify $x^*(i)$, then a superior least cost solution $x(i)$, excluding x' and x'', may be selected instead. Early termination becomes possible (though is not compulsory) upon encountering an $x^*(i)$ that yields $f(x^*(i)) < min(f(x'),f(x''),f(x(p)))$, where $x(p)$ is the minimum cost $x(h)$ for all $h < i$. The procedure will continue if $x(i)$, in contrast to $x^*(i)$, yields a smaller $f(x)$ value than x' and x'', since $x(i)$ effectively adopts the role of x'.

2.1   Simultaneous Relinking

Figure 7-10 shows that the path relinking approach is applied starting from both ends, that is, using x' as the initiating solution and as x'' the guiding solution as well as using x'' as the initiating solution and x' as the guiding solution. The simultaneous relinking approach starts with both endpoints x' and x'' simultaneously producing two sequences $x' = x'(1)$, ..., $x'(r)$ and $x'' = x''(1)$, ..., $x''(s)$. The choices in this case are designed to yield $x'(r) = x''(s)$, for final values of r and s. To progress toward the point where $x'(r) = x''(s)$, the choice rules should be such that the x' path approaches the last solution in the current x'' path and the other way around. Figure 7-12 depicts the simultaneous relinking approach.

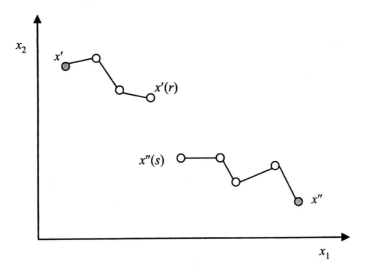

Figure 7-12. Illustration of simultaneous relinking

The simultaneous relinking may be viewed as a process for which two guiding solutions are dynamically changing until they converge to a single point. The next solution visited in the x' path illustrated in Figure 7-12 should be such that $x'(r)$ moves to $x'(r + 1)$ with the criterion of minimizing the number of moves remaining to reach $x''(s)$. Similarly, $x''(s)$ must move to $x''(s + 1)$ with the criterion of minimizing the number of moves remaining to reach $x'(r)$. From these options, the move is selected that produces the smallest $f(x)$ value, thus also determining which of r or s is incremented on the next step. Basing the relinking process on more than one neighborhood also produces a useful variation.

2.2 Dealing with Infeasibility

Strategic oscillation is a mechanism used in tabu search to allow the process to visit solutions around a "critical boundary", by approaching such a boundary from both sides. The most common application of strategic oscillation is in constrained problems, where the critical boundary is the feasibility boundary. The search process crosses the boundary from the feasible side to the infeasible side and also from the infeasible side to the feasible side. Glover and Kochenberger (1996) successfully applied strategic oscillation in the context of multidimensional knapsack problems.

Path relinking allows the search to cross the feasibility boundary by way of a *tunneling* strategy. The strategy is such that allows infeasible solutions to be visited while relinking x' and x''. It also allows for either x' or x'' to be infeasible but not both. Figure 7-13 shows the three cases of tunneling. The path depicted with a solid line visits infeasible solutions while going from x' to x'', both of which are feasible. The dotted path moves from an infeasible initiating solution to a feasible guiding solution, while the dashed path does the opposite. The dark dots in Figure 7-13 represent the guiding and the initiating solutions, while the white dots represent the solutions found during the relinking process.

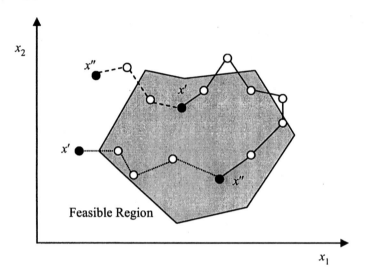

Figure 7-13. Three cases of tunneling

The tunneling strategy protects the search from becoming "lost" in the infeasible region, since feasibility evidently must be recovered by the time x'' is reached. When x'' is allowed to be infeasible, the relinking path may stop as soon as it leaves the feasible region or continue until reaching x'', since it is possible (although unlikely in some settings) for the path to go back to the feasible region before reaching x''. The tunneling effect therefore offers a chance to reach solutions that might otherwise be bypassed. If tunneling is combined with the simultaneous relinking approach at least one of $x'(r)$ and $x''(s)$ may be kept feasible.

To achieve a balance between intensification and diversification, it is appropriate to select the points x' and x'' by to a distance measure. Choosing x' and x'' close to each other stimulates intensification, while choosing them to maximize the distance between them stimulates diversification.

2.3   Extrapolated Relinking

The path relinking approach goes beyond consideration of points "between" x' and x'' in the same way that linear combinations extend beyond points that are expressed as convex combinations of two endpoints. In seeking a path that continues beyond x'' (starting from the point x') we invoke a tabu search concept that forbids adding tabu-active attributes back to the current solution. Consider, for example the initiating and guiding solutions x' and x'' for a binary problem in Figure 7-14. The sequence of relinking steps show the transformation of x' into x'' and the extrapolated path consisting of $x''(1)$ and $x''(2)$. The list of forbidden moves in Figure 7-14 allows the process to continue after reaching x'' without reversing some of the previous moves. The extrapolated path could be made longer if upon reaching $x''(2)$, the list of forbidden moves is reduced to contain only those moves made after reaching x'' (i.e., $x_5 \rightarrow 1$ and $x_2 \rightarrow 1$).

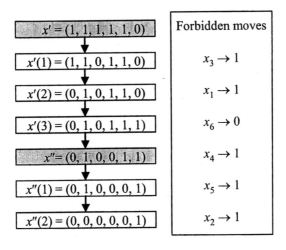

Figure 7-14. Illustration of extrapolated relinking

The illustration in Figure 7-14 may be generalized as follows. Let $A(x)$ denote the set of solution attributes associated with ('contained in') x, and let A_drop denote the set of solution attributes that are dropped by moves performed to reach the current solution $x'(i)$, starting from x'. (Such attributes may be components of the x vectors themselves, or may be related to these components by appropriately defined mappings.)

Define a *to-attribute* of a move to be an attribute of the solution produced by the move, but not an attribute of the solution that initiates the move. Similarly, define a *from-attribute* to be an attribute of the initiating solution but not of the new solution produced. Then we seek a move at each step to maximize the number of *to-attributes* that belong to $A(x'') - A(x'(i))$, and subject to this to minimize the number that belong to $A\_drop - A(x'')$. Such a rule generally can be implemented very efficiently by appropriate data structures.

In the illustration of Figure 7-14, a move consists of changing the value of a single variable from zero to one or from one to zero. That is, a variable x_i is selected and the new value is equal to $1 - x_i$. Therefore, the set $A(x'') - A(x')$ consists in all the variables for which $abs(x'_i - x''_i) = 1$, that is x_1, x_3, x_4, and x_6.

Once $x(r) = x''$ is reached, the process continues by modifying the choice rule as follows. The criterion now selects a move to maximize the number of its *to-attributes* not in *A_drop* minus the number of its *to-attributes* that are in *A_drop*, and subject to this to minimize the number of its *from-attributes* that belong to $A(x'')$. The *A_drop* at the time the relinking process reaches x'' in Figure 7-14, contains all the variables and corresponding values that appear in the "forbidden moves" list. In our illustration, the combined criteria are meaningful only if moves that change the value of more than one variable at a time are used. For example, the move $(x_1 \rightarrow 1, x_2 \rightarrow 1, x_5 \rightarrow 1)$ has one attribute $(x_1 = 1)$ in *A_drop* and two $(x_2 = 1, x_5 = 1)$ that are not. The resulting value for this combined move is one.

The combination of these criteria establishes an effect analogous to that achieved by the standard algebraic formula for extending a line segment beyond an endpoint. (The secondary minimization criterion is probably less important in this determination.) The path then stops whenever no choice remains that permits the maximization criterion to be positive. The maximization goals of these two criteria are of course approximate, and can be relaxed.

For neighborhoods that allow relatively unrestricted choices of moves, this approach yields a path extending beyond x'' that introduces new attributes, without reincorporating any old attributes, until no move remains that satisfies this condition. (This is what happens in Figure 7-14 when the entire list of forbidden moves is used after reaching x''.) The ability to go beyond the limiting points x' and x'' creates a form of diversification analogous to that provided by the original scatter search approach. At the same time the exterior points are influenced by the trajectory that links x' and x''.

2.4   Multiple Guiding Solutions

New points can be generated from multiple guiding solutions as follows. Instead of moving from a point x' to (or through) a second point x'', we replace x'' by a collection of solutions X''. Upon generating a point $x(i)$, the options for determining a next point $x(i + 1)$ are given by the union of the solutions in X'', or more precisely, by the union A'' of the attribute sets $A(x)$, for $x \in X''$. A'' takes the role of $A(x)$ in the attribute-based approach previously described, with the added stipulation that each attribute is counted (weighted) in accordance with the number of times it appears in elements $A(x)$ of the collection. Still more generally, we may assign a weight to $A(x)$, which thus translates into a sum of weights over A'' applicable to each attribute, creating an effect analogous to that of creating a weighted linear combination in Euclidean space. Parallel processing can be applied to operate on an entire collection of solutions $x' \in X'$ relative to a second collection $x'' \in X''$ by this approach. Further considerations that build on these ideas are detailed in Glover (1994b), but they go beyond the scope of our present development.

The path relinking with multiple guiding solutions generates new elements by a process that emulates the strategies of the original scatter search approach at a higher level of generalization. The reference to neighborhood spaces makes it possible to preserve desirable solution properties (such as complex feasibility conditions in scheduling and routing), without requiring artificial mechanisms to recover these properties in situations where they may otherwise become lost.

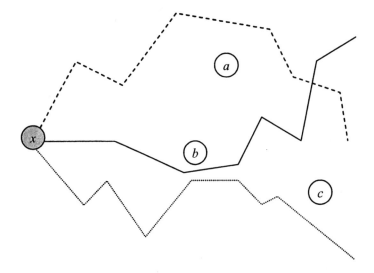

Figure 7-15. Neighborhood space paths with multi guiding solutions

Promising regions may be searched more thoroughly in path relinking by modifying the weights attached to attributes of guiding solutions, and by altering the bias associated with solution quality and selected solution features. Figure 7-15 depicts the type of variation that can result, where point x represents an initiating solution and points a, b, and c represent guiding solutions. For appropriate choices of the reference points (and neighborhoods for generating paths from them), principles such as those discussed in Glover and Laguna (1997) suggest that additional elite points are likely to be found in the regions traversed by the paths, upon launching new searches from high quality points on these paths.

2.5   Constructive Neighborhoods

A natural variation of path relinking occurs by using constructive neighborhoods for creating new trial solutions from a collection of initiating and guiding solutions. In this case the guiding solutions consist of subsets of elite solutions, as before, but the initiating solution begins as a partial (incomplete) solution or even as a null solution, where some of the components of the solutions, such as values for variables, are not yet assigned. The use of a constructive neighborhood permits such an initiating solution to "move toward" the guiding solutions, by a neighborhood path that progressively introduces elements contained in the guiding solutions, or that are evaluated as attractive based on the composition of the guiding solutions. The idea of using constructive (and destructive) neighborhood in the context of path relinking was originally described in Chapter 4 of Glover and Laguna (1997). In this section, we elaborate in that idea and include details from Glover, Laguna and Martí (2000).

The evaluations can be conceived as produced by a process where the guiding solutions *vote* for attributes to be included in the initiating solution. It is possible, for example, that a certain partial configuration may be reached where none of the attributes of the guiding solutions can be incorporated within the existing solution, relative to a given constructive neighborhood. Then it is important to still be able to select a next constructive step, by relying upon the voting process for evaluating moves. This same consideration can arise in transition neighborhoods, though it is encountered less frequently there.

Combinations created in this way are called *structured combinations*, and their generation rests upon three properties.

- *Representation property.* Each guiding solution represents a vector of votes for particular decisions (e.g., the decision of assigning a specific value to a particular variable).
- *Trial solution property.* The votes prescribed by a guiding solution translate into a trial solution to the problem of interest by a well-defined process (determined by the neighborhood structure).
- *Update property.* If a decision is made according to the votes of a given vector, a clearly defined rule exists to update all voting vectors for the residual problem so that Properties 1 and 2 continue to hold.

Features of these properties in particular contexts may be clarified as follows.

– *Elaboration of Property 1*: Standard solution vectors for many problems can directly operate as voting vectors, or can be expanded in a natural way to create such vectors. For instance, a solution vector for a job shop scheduling problem can be interpreted as a set of 0-1 votes for predecessor decisions in scheduling specific jobs on particular machines.
– *Elaboration of Property 2*: A set of "yes-no" votes for items to include in a knapsack, for instance, can be translated into a trial solution according to a designated sequence for processing the votes (such as determined by benefit-to-weight ratios), until either the knapsack is full or all votes are considered. More general numerical votes for the same problem may additionally prescribe the sequence to be employed, as where knapsack items are rearranged so the votes occur in descending order. (The voting vectors are not required to represent feasible solutions to the problems considered, or even represent solutions in a customary sense at all.)
– *Elaboration of Property 3*. Upon assigning a specific value to a particular variable, all votes for assigning different values to this variable effectively become cancelled. Property 3 then implies that the remaining updated votes of each vector retain the ability to be translated into a trial solution for the residual problem in which the assignment has been made.

Concrete illustrations of processes for generating structured combinations by reference to these properties are provided in Glover (1994b). These same kinds of processes can be implemented by reference to destructive neighborhoods—that is, neighborhoods that allow the removal of less attractive elements. Typically, destructive processes are applied to solutions that begin with an "excessive assignment" (such as too many elements to satisfy cardinality or capacity restrictions).

2.6 Vocabulary Building

Vocabulary building creates structured combinations not only by using the primitive elements of customary neighborhoods, but also building and joining more complex assemblies of such elements. The process receives its name by analogy with the process of building words progressively into useful phrases, sentences and paragraphs, where valuable constructions at each level can be visualized as represented by "higher order words," just as natural languages generate new words to take the place of collections of words that embody useful concepts.

The motive underlying vocabulary building is to take advantage of those contexts where certain partial configurations of solutions often occur as components of good complete solutions. A strategy of seeking "good partial configurations"—good vocabulary elements—can help to circumvent the combinatorial explosion that potentially results by manipulating only the most primitive elements by themselves. The process also avoids the need to reinvent (or rediscover) the structure of a partial configuration as a basis for building a good complete solution. (The same principle operates in mathematical analysis generally, where basic premises are organized to produce useful lemmas, which in turn facilitate the generation of more complex theorems.)

Laguna, et al. (2000) implemented a simple vocabulary building process to extract the basic structure of reference solutions in a scatter search procedure. This implementation was in the context of scheduling parallel resources used to process a set of jobs, some of which may be interrupted by an uncertain amount of time. The resources may be machines in a production system, computers with specialized software packages (as those needed for engineering designs), or highly specialized technicians. Suppose that in a 20-job problem with 3 parallel machines, the two reference solutions shown in Table 7-7 are to be combined. (A complete solution to the problem includes the assignment of jobs to machines, the sequence of the deterministic jobs in each machine and the interruption policy for the jobs with stochastic interruption. For the purpose of this illustration, however, we focus on the job assignments only.)

Table 7-7. Job assignments in two reference solutions

Solutions	Machines		
	1	2	3
1	1, 3, 6, 9, 12, 17, 18	2, 5, 8, 10, 11, 19, 20	4, 7, 13, 14, 15, 16
2	1, 4, 5, 10, 11, 20	2, 3, 6, 12, 13, 15	7, 8, 9, 14, 16, 17, 18, 19

The vocabulary building mechanism assigns jobs to machines based on identifying job groups within two reference solutions. The identification of these groups is considered the extraction of a "basic" structure embedded in the two reference solutions. This basic structure is preserved in any new trial solution generated from the given pair of reference solutions. Since the basic structure assigns only a subset of all jobs, the remaining jobs are assigned to machines following a deterministic process based on votes. So, the entire process of assigning jobs to machines to create new trial solutions includes the use of a constructive neighborhood of the type described in the previous section.

From the reference solutions in Table 7-7, the vocabulary building process in Laguna, et al. (2000) solves an assignment problem that finds the best matching of machines. The objective function maximizes the total number of jobs that appear in the matched machines. For example, if we math machine 1 from solution 1 with machine 1 from solution2 in Table 7-7, we obtain only one common job (i.e., job number 1). So, the assignment problem associated with the solutions in Table 7-7 is as follows:

Maximize $\quad x_{11} + 3\,x_{12} + 3\,x_{13} + 4\,x_{21} + x_{22} + 2\,x_{23} + x_{31} + 2\,x_{32} + 3\,x_{33}$

subject to $\quad x_{11} + x_{12} + x_{13} = 1$
$$x_{21} + x_{22} + x_{23} = 1$$
$$x_{31} + x_{32} + x_{33} = 1$$

$$x_{ij} \in \{0,1\} \qquad i, j = 1, 2, 3.$$

In this formulation, $x_{ij} = 1$ if machine i from reference solution 1 is matched with machine j of reference solution 2. The optimal solution to this problem is $x_{12} = x_{21} = x_{33} = 1$ and all other variables equal to zero. Therefore the basic structure of the solutions generated by the two reference solutions in Table 7-7 has the partial assignment showed in Table 7-8.

Table 7-8. Partial assignment of jobs to machines

Machines		
1	2	3
3, 6, 12	5, 10, 11, 20	7, 14, 16

An extension of this basic design would consist of applying a generalized procedure for finding partial assignments in more than two reference solutions. In other words, the vocabulary building process would attempt to find the common elements in elite solutions to use those partial assignments as the basis for creating additional trial solutions.

Vocabulary building has an additional useful feature in some problem settings by providing compound elements linked by special neighborhoods that are more exploitable than the neighborhoods that operate on the primitive elements. For example, a vocabulary-building proposal of Glover (1992) discloses that certain subassemblies (partial "tours") for traveling salesman problems can be linked by exact algorithms to produce optimal unions of these components. Variants of this strategy have more recently been introduced by Aggarwal, Orlin and Tai (1997) as a proposal for modifying traditional genetic algorithms, and have also been applied to weighted clique problems by Balas and Niehaus (1998). A particularly interesting application occurs in the work of Lourenço, Paixao and Portugal (2001), who use such concepts to create "perfect offspring" by solving a set-covering problem in the process of combining solutions to a bus driver scheduling problem.

In general, vocabulary building relies on destructive as well as constructive processes to generate desirable partial solutions, as in the early proposals for exploiting strongly determined and consistent variables— which essentially "break apart" good solutions to extract good component assignments, and then subject these assignments to heuristics to rebuild them into complete solutions. Construction and destruction therefore operate hand in hand in these approaches. A graphical representation of vocabulary building is depicted in Figure 7-16.

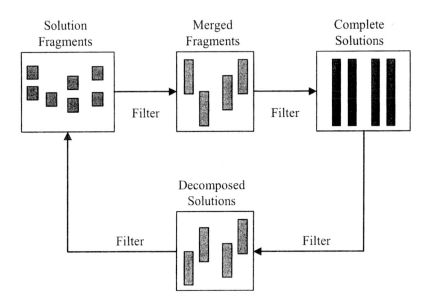

Figure 7-16. Vocabulary building process

In vocabulary building, solution fragments are merged using heuristic or exact procedures. This is illustrated as the application of a "filter" between the "Solution fragments" box and the "Merged Fragments" box in Figure 7-16. For example, in a traveling salesperson problem, the solution fragments may be edges that are merged using a filter (heuristic or exact) to form tour segments. An additional filter is then applied to turn the tour segments into complete solutions. The destruction phase of vocabulary building applies a filter to decompose solutions and an additional filter to identify the relevant fragments from the decomposed solutions. In the traveling salesperson example, a complete tour would be subjected to a filter to identify relevant tour segments and then another filter to extract smaller fragments, which may be either smaller tour segments or individual edges.

The vocabulary building processes in Laguna, et al. (2000) and Lourenço, Paixao and Portugal (2001) are simplified versions of the process depicted in Figure 7-16. Those implementations use complete (reference) solutions and two filters to generate new trial solutions. The first filter extracts solution fragments and the second filter constructs a new trial solution based on the fragments. Both use exact procedures as the first filter and heuristics for the second filter. A schematic representation of this simplified process is depicted in Figure 7-17.

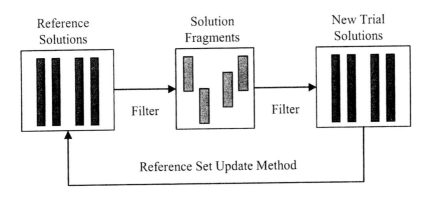

Figure 7-17. Simplified vocabulary building process

Although we use the scatter search nomenclature in Figure 7-17, the vocabulary building process may be applied in connection with other evolutionary approaches as done by Lourenço, Paixao and Portugal (2001) in their genetic algorithm. In such a case, the reference solutions are parents from the current population and the new trial solutions are the offspring. Regardless of the way the reference solutions are obtained, the filters that extract fragments from these solutions and generate new solutions from those fragments are mechanisms that can be directly applied to variety of search procedures.

2.7   Computer code

We now consider a basic path relinking implementation for the linear ordering problem. The code is included in the Chapter7 folder of the accompanying disc. The procedure has many common elements with the scatter search implementation for the LOP described in Chapter 4. The path relinking procedure starts with the creation of an initial reference set of solutions (*RefSet*). The Diversification Generation Method of Chapter 4 is used to build a set *P* of diverse solutions. The initial reference set is built with the Reference Set Update Method of Chapter 4 and the solutions in

RefSet are ordered according to quality, where the best solution is the first one in the list. Instead of applying the Combination Method of Chapter 4, a path relinking method is applied to all solutions pairs in *RefSet*. For each pair (x', x'') two paths are initiated; one from x' to x'' and the other from x'' to x'. Function 7-1 implements the search from the initiating solution `sol1` to the guiding solution `sol2`, which is called one time with `sol1` = x' and `sol2` = x'' and a second time with `sol1` = x'' and `sol2` = x'. This function returns the best solution found in the `newsol` array.

As in the scatter search designs of the tutorial chapters, we use the so-called *static update* of the reference set after the application of the path relinking method. Trial solutions that result from the path relinking method (`newsol`) are placed in a solution *Pool*. After the application of both the Combination Method and the Improvement Method, *Pool* is full and the reference set is updated. The new reference set consists of the best *b* solutions from the solutions in the current reference set and the solutions in *Pool*. As in SS, if the reference set remains unchanged after the updating procedure, a rebuilding step is performed; otherwise, the reference set is used for a new round of path relinking steps.

Function 7-1, `SSPath_Relinking`, starts by creating the `csol` array to store the current (intermediate) solution in the relinking from `sol1` to `sol2`. Solution `csol` is initially equal to `sol1`. Then the function computes the `csol_inv` inverse solution of `csol`, where the inverse of a solution in this context is defined as `sol_inv[j]` = `i` if and only if `sol[i]` = `j`. The inverse solution `sol2_inv` of `sol2` is also computed. The function selects a sector `a` and determines both its current position `pos` in `csol` (`pos` = `csol_inv[a]`) and its new position `new_pos` according to `sol2` (`new_pos` = `sol2_inv[a]`). The function scans the sectors in the order given by their measure of attractiveness, which in this case is the sum of the weights in their corresponding row (`pb->row_sum`). This order is established with a call to the `SSOrder_i` function. At each relinking step, we test if the current solution improves upon the best solution found in order to update `newsol` if necessary. The relinking steps terminate when `csol` becomes `sol2`.

The `SSImprove_solution` function in Section 3.1 of Chapter 4 is used to improve some of the solutions generated during the relinking phase. This Improvement Method is applied every `NumImp` steps of the relinking process. In this implementation, this value has been set to 5. The `SSImprove_solution` function is called from a temporary solution `temp_sol` where the intermediate solution `csol` has been previously copied. `SSImprove_solution` overwrites `temp_sol` with the local optimum found. If the solution returned by `SSImprove_solution` is better than the best solution found the `newsol` solution is updated.

Function 7-1. SSPath_Relinking — File: PRRefSet.c

```
void SSPath_Relinking (SS *pb, int sol1[], int sol2[],
                        int value1 ,int newsol[])
{
    int i,j,a,pos,*order,new_pos,NumImp=5;
    int best_value,*temp_sol,temp_value;
    int *csol,cvalue;      /* Current (intermediate) solution */
    int *csol_inv,*sol2_inv; /* sol[i]=j <--> sol_inv[j]=i */

    /* Initialize current and best sol */
    temp_sol = SSInt_array(pb->nvar);
    csol = SSInt_array(pb->nvar);
    for(j=1;j<=pb->nvar;j++)
      csol[j]=newsol[j]=sol1[j];
    cvalue=best_value=value1;

    /* Compute the inv. order  */
    sol2_inv  = SSInt_array(pb->nvar);
    for(j=1;j<=pb->nvar;j++)
      sol2_inv[sol2[j]]=j;
    csol_inv  = SSInt_array(pb->nvar);
    for(j=1;j<=pb->nvar;j++)
      csol_inv[csol[j]]=j;

    /* Order sectors according to row_sum */
    order = SSOrder_i(pb->row_sum,pb->nvar,1);

    /* Construct a path from sol1 to sol2 */
    while(!SSEqualSol(csol,sol2,pb->nvar))
    {
      for(i=1;i<=pb->nvar;i++)
      {
        a=order[i];
        pos=csol_inv[a];
        new_pos=sol2_inv[a];
        cvalue += SSCompute_Insert(pb,csol,pos,new_pos);
        SSPerform_Move(csol,csol_inv,pos,new_pos);

        /* Update best solution found */
        if(cvalue > best_value)
        {
          best_value=cvalue;
```

```
        for(j=1;j<=pb->nvar;j++)
          newsol[j]=csol[j];
      }

      /* Apply a local search improvement */
      if(i % NumImp ==0)
      {
        temp_value=cvalue;
        for(j=1;j<=pb->nvar;j++)
          temp_sol[j]=csol[j];
        SSImprove_solution(pb,temp_sol,&temp_value);

        /* Update the best solution found */
        if(temp_value > best_value)
        {
          best_value=temp_value;
          for(j=1;j<=pb->nvar;j++)
            newsol[j]=temp_sol[j];
} } } }
    free(order+1);free(csol+1);
    free(temp_sol+1);free(sol2_inv+1);
}
```

---

3.   INTENSIFICATION AND DIVERSIFICATION

A significant feature that distinguishes scatter search and path relinking from other evolutionary approaches is that intensification and diversification processes are not conceived to be embedded solely within the mechanisms for combining solutions, or within supplementary "mutation" strategies based on randomly varying reference solutions to produce new trial solutions.[1]

The initial connections between scatter search and strategies involving measures of frequency and influence are related to the exploitation of *consistent* and *strongly determined* variables. Loosely speaking, consistent

---

[1] Within the last few years, some researchers in the evolutionary computation field have begun to adopt aspects of scatter search and path relinking by incorporating systematic strategies for achieving intensification and diversification, instead of relying on randomization to achieve less purposeful forms of variation. However, some of the latest literature still disallows this type of approach as a legitimate feature of evolutionary computation. For example, Fogel (1998) says that the main disciplines of evolutionary computation all involve a process whereby "New solutions are created by randomly varying the existing solutions."

variables are those more frequently found in reference solutions, while strongly determined variables are those that would cause the greatest disruption by changing their values. The approach based on this idea is to isolate the variables that qualify as more consistent and strongly determined and then to generate new trial solutions that give these variables their "preferred values." For instance, in the process of solving a 0-1 knapsack problem some of the variables will typically have highly profitable profit-weight ratios and others will have highly unprofitable ratios. The variables with ratios at the extremes qualify as being strongly determined, since they are nearly compelled to assume particular values. In the same context, a consistent variable is one that is frequently selected (i.e., is given a value of one) even though its profit-weight ratio is not one of the most profitable ones. Hence, a consistent variable is one that is strongly determined at a particular value (Glover, 1977).

The strategy that exploits the notion of consistent and strongly determined variables naturally falls in the category of search intensification, in the sense that it attempts to take advantage of features associated with good solutions. It is predicated on a highly explicit analysis of the frequency by which attributes belong to high quality solutions. This stands in notable contrast to the philosophy of other mainstream evolutionary procedures, where the relevance of attribute membership in solutions is left to be guessed mainly by the device of randomly shuffling and combining solutions.

The approach called *strategic oscillation,* briefly discussed at the beginning of Section 2.2 of this Chapter, that was introduced with the original scatter search proposal and has been more commonly used in connection with tabu search, is important for linking intensification and diversification. The basic idea of this approach is to identify critical regions of search, and to induce the search pattern to visit these regions to various depths within their boundaries, by a variable pattern that approaches and retreats from the boundaries in oscillating waves. Glover, Laguna and Martí (2000) show several examples of oscillation regions and their corresponding boundaries. However, the most common form of strategic oscillation relates to the feasible and infeasible regions which are delimited by the feasibility boundary. The strategic oscillation approach operates by moving through one region to approach the boundary, and then either crosses the boundary or reverses direction to move back into the region just traversed. This process is illustrated in Figure 7-18, where the search moves in and out of a two-dimensional feasible region. The oscillation in Figure 7-18 is not symmetrical with respect to the number of moves performed in each side of the oscillation boundary, illustrating the fact that the oscillation pattern may be controlled with a variety of rules based on mechanisms such as infeasibility thresholds or short-term memory functions.

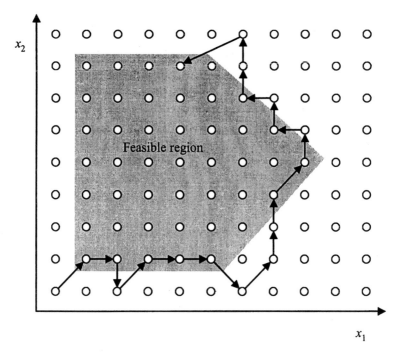

Figure 7-18. Strategic oscillation illustration

The Improvement Method for 0-1 knapsack problems that we described in Chapter 3 could be extended to include a strategic oscillation component. Starting from an infeasible solution, the Improvement Method makes moves to "take items out of the knapsack" until the total weight is not violated. This means that variables with the value of 1 in the current solution are switched to zero until the total weight is less than or equal to the allowed weight. When the solution becomes feasible, the direction is reversed and items are added to the knapsack as long as the total weight is not exceeded. In the implementation of Chapter 3, the process stops after reaching a point in which no items can be added without violating the capacity constraint.

A simple oscillation strategy would be to allow adding up to n_a items after crossing the boundary in the infeasible direction and deleting up to n_d items after crossing the boundary in the feasible direction. If the rules for adding and deleting items are properly chosen and a simple short-term memory mechanism is used, the Improvement Method can explore beyond the first local optimum and examine the promising solutions that lie close to the feasibility boundary. The values of n_a and n_d need not be equal, creating an asymmetrical oscillation around the feasibility boundary. In general, however, it seems appropriate to keep both of these values small with respect to the total number of items in the problem.

This is the approach taken by Glover and Kochenberger (1996) in the context of multidimensional knapsack problems. As the authors eloquently describe in their article, they focus on a subset of tabu search elements to create a search process based on a flexible memory structure that is updated at critical events. The tabu status of a potential move is determined through the integration of regency-based and frequency-based memory information (see Chapter 6). A balance between intensification and diversification is accomplished by a strategic oscillation scheme that probes systematically to varied depths on each side of the feasibility boundary. These oscillations, coupled with dynamic tabu information, guide the search process toward different critical events. We believe that research regarding strategic oscillation and the intensification-diversification dichotomy deserves further consideration.

Chapter 8

SCATTER SEARCH APPLICATIONS

We thought that we had the answers, it was the questions we had wrong.

Bono, U2

This section provides a collection of "vignettes" that briefly summarize applications of scatter search (SS) and path relinking (PR) in a variety of settings. These vignettes are edited versions of reports by researchers and practitioners who are responsible for the applications. A debt of gratitude is owed to the individuals whose contributions have made this summary possible.

1. NEURAL NETWORK TRAINING

The problem of training a neural network (NN) is that of finding a set of weights w that minimizes an error measure. In neural networks used for classification, the error relates to the number of misclassified items. In a neural network used for estimation, the error relates to the difference between the estimated value and the actual value. Laguna and Martí (2001b) developed a scatter search procedure for training a neural network in the context of optimizing simulations. The training procedure was applied to a feedforward network with a single hidden layer and the objective was to minimize the mean squared error (MSE). The authors assumed that there were n decision variables in an optimization-simulation problem and that the neural network had m hidden neurons with a bias term in each hidden neuron and an output neuron. A schematic representation of the network appears in Figure 8-1. Note that the weights in the network are numbered sequentially starting with the first input to the first hidden neuron. Therefore, the weights for all the inputs to the first hidden neuron are w_1 to w_n. The bias term for the first hidden neuron is w_{n+1}.

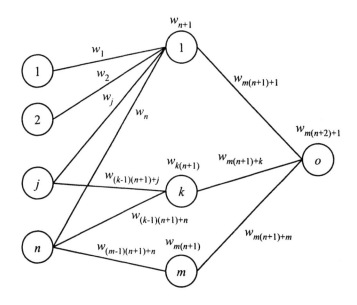

Figure 8-1. Neural network with one hidden layer and one output

The procedure starts with data normalization and the initialization of the reference set (*RefSet*), which consists of b solutions. The initialization is done by first constructing a set P of *PSize* solutions (bounded between *wlow* and *whigh*) with the Diversification Generation Method. The method is based on a controlled randomization scheme developed by Glover, Laguna and Martí (2000). *RefSet* is populated with the best $b/2$ solutions in P to which the Improvement Method is applied. *RefSet* is then completed with $b/2$ more solutions that are perturbations of the first $b/2$. The perturbation consists of multiplying each weight by $1 + U[-0.05, 0.05]$, where U is the uniform distribution.

As customary, the solutions in *RefSet* are ordered according to quality, where the best solution is the first one in the list. The Combination Method is applied to all new solution pairs. A pair is new if at least one of the solutions in the pair is new. The Combination Method consists of linear combinations of weights. The Improvement Method is applied to the best b trial solutions generated during the combination step. If an improved trial solution improves upon the worst solution currently in *RefSet*, the improved trial solution replaces the worst and *RefSet* is reordered.

The procedure is designed to intensify the search around the best-known solution. At each intensification iteration, the best-known solution is perturbed (multiplied by 1 + U[-0.05,0.05]) and the Improvement Method is applied. The best solution is updated if the perturbation plus the improvement generates a better solution. The intensification phase is abandoned after a specified number of iterations elapsed without improving the best solution. The training procedure stops when the number of objective function evaluations reaches an allowed total. After experimentation, the parameters *wlow, whigh, b* and *IntLimit* were set to –2, 2, 10 and 20 respectively.

The SS procedure employs the well known Nelder and Mead (1965) simplex optimizer as its Improvement Method. The downhill simplex method is a non-linear unconstrained optimizer with nothing in common with the simplex method of linear programming. Given a set of weights *w*, the simplex method starts by perturbing each weight to create an initial simplex from which to begin the local search, which basically consists of movements on the vertices of this simplex.

The computational experiments compare the scatter search implementation with the best known methods and show that SS reaches a prediction accuracy that is sufficient for the purpose of filtering out potentially bad solutions generated during the optimization of a simulation, and it does so within a computational time that is practical for on-line training.

1.1   Computer code

One of the most innovative aspects of this scatter search implementation is its Reference Set Update Method. Since the objective value (i.e., MSE) is extremely sensitive to changes in the values of the variables, the notion of "adding diversity to the *RefSet*" must be reinterpreted in this context and this is reflected in the implementation of the Reference Set Update Method. Moreover, the experiments reveal that the use of a local optimizer, although computationally expensive, enhances the search significantly. Hence, another innovative feature in this implementation is the use of local search during the updating of the reference set. Function 8-1, `Update_Refset`, implements the Reference Set Update Method. First, the best *b*/2 elements of *RefSet* are submitted to the `SSimprove_solution` function, which executes the simplex method as long as the current solution is improved. The remaining solutions in the *RefSet* are removed and replaced with a perturbation of the first *b*/2. The perturbation is achieved by adding the value of the `perturb` variable to each solution. This small variation adds enough diversity to the *RefSet* and enables the search to continue. The Chapter8 folder in the accompanying compact disc contains the complete

code of this scatter search implementation.   Five different examples are
included in the file data.c to test the neural network training procedure.

Function 8-1. Update_Refset — File: SS_RefSet.c

```
void Update_RefSet(Net *p,SS *prob)
{
    double *current,*min_dist,*value,**solutions;
    int j,i,k,*index2;
    double perturb;

    prob->iter++;
    prob->digits++;
    /* Memory allocation */
    current = SSallocate_double_array(prob->n_var);
    if(!current) SSabort("Memory allocation problem");
    min_dist = SSallocate_double_array(prob->PSize);
    if(!min_dist) SSabort("Memory allocation problem");
    value = SSallocate_double_array(prob->PSize);
    if(!min_dist) SSabort("Memory allocation problem");
    solutions=SSallocate_double_matrix(prob->PSize,
            prob->n_var);
    if(!solutions) SSabort("Memory allocation problem");

    /* Improve the first b/2 elements */
    if(prob->LS ){
      prob->ImpCount=0;
      for(i=1;i<=prob->b1/2;i++)
        SSimprove_solution(p,prob->RefSet1[prob->order1[i]],
                &(prob->value1[prob->order1[i]]));
    }

    /*Add the second b/2 to RefSet */
    for(i=1;i<=prob->b1/2;i++)
    {
      k = prob->order1[i+(prob->b1/2)];
      for(j=1;j<=prob->n_var;j++)
      {
        perturb = prob->RefSet1[i][j]*
                (-0.01+0.02*(rand()/(double)RAND_MAX));
        if(fabs(perturb) < EPSILON)
                perturb = -0.1+0.2*(rand()/(double)RAND_MAX);
        prob->RefSet1[k][j] = prob->RefSet1[i][j] + perturb;
```

```
   }
   prob->value1[k]= compute_error(p,prob->RefSet1[k]);

   if(prob->LS)    SSimprove_solution(p,prob->RefSet1[k],
                   &(prob->value1[k])));
   prob->iter1[k]  = 1;
}


index2 = orden_indices(prob->value1,prob->b1,-1);

for(i=1;i<=prob->b1;i++)
{
   prob->order1[i] = index2[i];
   prob->iter1[i]  = prob->iter;
}
prob->new_elements = 1;
free(index2);free(current);free(min_dist);free(value);
SSfree_double_matrix(solutions,prob->PSize);
```
}

2.   MULTI-OBJECTIVE BUS ROUTING

Corberán et al. (2002) address the problem of routing school buses in a rural area. The authors approach this problem with a node routing model with multiple objectives that arise from conflicting viewpoints. From the point of view of cost, it is desirable to minimize the number of buses (m) used to transport students from their homes to school and back. And from the point of view of service, it is desirable to minimize the time that a given student spends in route (let *tmax* be the maximum time in the bus). Most of the literature deals primarily with single-objective problems and the models with multiple objectives typically employ a weighted function to combine the objectives into a single one.

Corberán et al. (2002) developed a solution procedure that considers each objective separately and search for a set of efficient solutions instead of a single optimum. The solution procedure is based on constructing, improving and then combining solutions within the SS framework. The method consists of the following elements:

H1 and H2:   Two constructive heuristics to generate routes
SWAP: An exchange procedure to find a local optimal value for the length of each route

INSERT: An exchange procedure to improve upon the value of
                 tmax
COMBINE: A mechanism to combine solutions in a reference set of
                 solutions in order to generate new ones

---

1. *Construct solutions* — Apply constructions heuristics H1 and H2 with
    several values of *tmax*.
2. *Improve solutions* — Apply SWAP to each route in a solution and
    INSERT to the entire solution. Finally, apply SWAP o any route
    changed during the application of INSERT.
3. *Build solution pools* — Put all solutions with the same number of routes
    in the same pool.
for (each solution pool) **do**
    4. *Build the reference set* — Choose the best *b* solutions in the pool to
        build the initial *RefSet*.
    while (new solutions in *RefSet*) **do**
        5. *Combine solutions* — Generate all the combined solutions from
            pairs of reference solutions where at least one solution in the
            pair is new.
        6. *Improve solutions* — Apply SWAP to each route in a solution
            and INSERT to the entire solution. Finally, apply SWAP to
            any route changed during the application of INSERT.
        7. *Update reference set* — Choose the best *b* solutions from the
            union of the current reference set and the improved trial
            solutions to update the *RefSet*.
    end while
end for

Figure 8-2. SS for multi-objective vehicle routing

The overall procedure operates as follows and is outlined in Figure 8-2.
The constructive heuristics H1 and H2 are applied with several values for
tmax and the resulting solutions are stored in separate pools, one for each
value of *m*. The larger the value of *tmax* the larger the frequency in which
the heuristics construct solutions with a small number of routes. Conversely,
solutions with large number of routes are obtained when the value of *tmax* is
decreased. The procedure then attempts to improve upon the solutions
constructed by H1 and H2. The improvement consists of first applying
SWAP to each route and then applying INSERT to the entire solution. If
any route is changed during the application of INSERT then SWAP is
applied one more time to all the changed routes. The procedure iterates
within a main loop, in which a search is launched for solutions with a
common number of routes. The main loop terminates when all the *m*-values
have been explored.

From all the solutions with m routes, the b best are chosen to initialize the reference set (*RefSet*). The criterion for ranking the solutions at this step is *tmax*, since all solutions have the same number of routes. The procedure performs iterations in an inner-loop that consists of searching for a solution with m routes with an improved *tmax* value. The combination procedure COMBINE is applied to all pairs of solutions in the current *RefSet*. The combined solutions are improved in the same way as described above, that is, by applying SWAP then INSERT and finally SWAP to the routes that changed during the application of INSERT. The improved trial solutions are placed in *ImpSet*. The reference set is then updated by selecting the best b solutions from the union of *RefSet* and *ImpSet*. Steps 5, 6 and 7 in the outline of Figure 8-2 are performed as long as at least one new solution is admitted in the reference set.

Note that after the reference set is updated, the combination procedure may be applied to the same solution pairs more than once. However, the combination procedure includes some randomized elements and therefore the combination of two solutions may result in a different outcome every time COMBINE is applied. Also, the size of the reference set is increased if the updating procedure fails to add at least one new solution. The additional solutions come from the original pool of solutions generated with the construction heuristics. The reference set size is increased up to $2*b$, where b is the initial size.

The computational testing showed that the SS implementation was capable of approximating the efficient frontier of this routing problem. Decision makers may use efficient solutions to find the best service level (given by the maximum route length) that can be obtained with each level of investment (given by the number of buses used). The results show that several of the solutions implemented in practice were not efficient.

3.   ARC CROSSING MINIMIZATION IN GRAPHS

Researches in the graph-drawing field have proposed several aesthetic criteria that attempt to capture the meaning of a "good" map of a graph. Although readability may depend on the context and the map's user, most authors agree that crossing reduction is a fundamental aesthetic criterion in graph drawing. In the context of a 2-layer graph and straight edges, the bipartite drawing problem or BDP consists of ordering the vertices in order to minimize the number of crossings.

A bipartite graph $G=(V,E)$ is a simple directed graph where the set of vertices V is partitioned into two subsets, V_1 (the left layer) and V_2 (the right layer) and where $E \subseteq V_1 \times V_2$. Note that the direction of the arcs has no effect on crossings so G is considered to be an undirected graph denoted by the triple (V_1, V_2, E), where E is the set of edges. Let $n_1 = |V_1|$, $n_2 = |V_2|$, $m = |E|$, and let $N(v) = \{w \in V \mid e = \{v, w\} \in E\}$ denote the set of neighbors of $v \in V$. A solution is completely specified by a permutation π_1 of V_1 and a permutation π_2 of V_2, where $\pi_1(v)$ or $\pi_2(v)$ is the position of v in its corresponding layer.

Laguna and Martí (1999) propose a path relinking (PR) procedure for arc crossing minimization within the GRASP framework. This is the first implementation that combines PR and GRASP, although Fleurent and Glover (1997) had proposed the use of elite solutions in connection with GRASP. In the proposed implementation of path relinking, the procedure stores a small set of high quality (elite) solutions to be used for guiding purposes. Specifically, after each GRASP iteration, the resulting solution is compared to the best three solutions found during the search. If the new solution is better than any one in the elite set, the set is updated. Instead of using attributes of all the elite solutions for guiding purposes, one of the elite solutions is randomly selected to serve as a guiding solution during the relinking process. The relinking in this context consists of finding a path between a solution found after an improvement phase and the chosen elite solution. Therefore, the relinking concept has a different interpretation within GRASP, since the solutions found from one GRASP iteration to the next are not linked by a sequence of moves (as in the case of tabu search). The relinking process implemented by Laguna and Martí (1999) may be summarized as follows.

The set of elite solutions is constructed during the first three GRASP iterations. Starting with the fourth GRASP iteration, every solution after the improvement phase is used as an initiating solution and is subjected to a relinking process that performs moves that transforms it into a guiding solution (i.e., an elite solution selected at random). The transformation is relatively simple, at each step, a vertex v is chosen from the initiating solution and is placed in the position occupied by this vertex in the guiding solution. So, if $\pi_1^g(v)$ is the position of vertex v in the guiding solution, then the assignment $\pi_1^i(v) = \pi_1^g(v)$ is made. We assume that an updating of the positions of vertices in V_1 of the initiating solution occurs. After this is done, an expanded neighborhood from the current solution defined by $\pi_1^i(v)$ and $\pi_2^i(v)$ is examined. The expanded neighborhood consists of a sequence of position exchanges of vertices that are one position away from each other, which are performed until no more improvement (with respect to crossing minimization) can be achieved. Once the expanded neighborhood has been explored, the relinking continues from the solution defined by $\pi_1^i(v)$ and $\pi_2^i(v)$ before the exchanges were made. The relinking finishes when the initiating solution matches the guiding solution, which will occur after $n_1 + n_2$ relinking steps.

Note that two consecutive solutions after a relinking step differ only in the position of two vertices (after the assignment $\pi_1^i(v) = \pi_1^g(v)$ is made). Therefore, it is not efficient to apply the expanded neighborhood exploration (i.e., the exchange mechanism) at every step of the relinking process. The parameter β is introduced to control the application of the exchange mechanism. In particular, the exchange mechanism is applied every β steps of the relinking process.

Overall, experiments with 3,200 graphs were performed to assess the merit of the procedure. The performance of the GRASP and path relinking implementations were compared with two methods: the iterated barycenter (Eades and Kelly, 1986) and a version of the tabu search algorithm (Martí, 1998). The former is the best of the simple heuristics for the BDP (Martí and Laguna, 1997), while the later has been proven to consistently provide the best solutions in terms of quality. The proposed GRASP-PR method is shown competitive in the set of problem instances for which the optimal solutions are known. For a set of sparse instances, the method performed remarkably well (outperforming the best procedures reported in the literature). The enhancements produced by path relinking suggest the potential merit of joining the PR guidance strategies with other multistart methods.

4. MAXIMUM CLIQUE

The Maximum Clique Problem (MCP) can be defined as follows. Given an undirected graph $G=(V, A)$ and $A(v_i)$ denoting the set of vertices v_j such that $(v_i, v_j) \in A$, then a graph $G_1=(V_1, A_1)$ is called a subgraph of G if $V_1 \subseteq V$, and for every $v_j \in V_1$, $A_1(v_i) = A(v_i) \cap V_1$. A graph G_1 is said to be complete if there is an arc for each pair of vertices. A complete subgraph is also called a clique. A clique is maximal, if it is not contained in any other clique. In the MCP the objective is to find a complete subgraph of largest cardinality in a graph. The clique number is equal to the cardinality of the largest clique of G.

The MCP is an important problem in combinatorial optimization with many applications which include: market analysis, project selection, and signal transmission. The interest for this problem led to the algorithm thread challenge on experimental analysis and algorithm performance promoted by the Second DIMACS Implementation Challenge (Johnson and Trick, 1996).

Cavique, Rego and Themido (2001) developed an experimental study for solving the Maximum Clique Problem using a scatter search framework.

The proposed algorithm considers structured solution combinations, weighted by a "filtering vector" playing the role of linear combinations. For the heuristic improvement a simple tabu search procedure based on appropriate neighborhood structures is used. Some special features and techniques have been introduced in this implementation.

Diversification Generation Method

The aim of the Diversification Generation Method is to create a set of solutions as scattered as possible within the solution space while also using as many variables (or solution attributes) as possible. In the MCP, all vertices in the graph G should be present in the solutions in *RefSet*.

When the algorithm starts, *RefSet* is initialized with a set of diverse solutions obtained by a constructive procedure, which starting from a single vertex, it adds at each step a new vertex to the current clique until a maximal clique is found. Starting from a different vertex not yet included in *RefSet*, the procedure is repeated as many times as the cardinality of the reference set. The clique value is used to order the solutions in *RefSet*.

Improvement Method

The improvement method has two phases: given a solution that is typically infeasible, the method first undertakes to recover feasibility; and afterward it attempts to increase the objective function value. Neighborhood structures based on add, drop, and node swap moves are used in the local search. The method handles infeasible solutions by implementing a strategic oscillation around the feasibility boundary.

Reference Set Update Method

This method must be carefully configured to mix diverse and high quality solutions and avoid the phenomenon of premature convergence of *RefSet*, which occurs when all the solutions are similar. The reference set is divided into two groups: the set of best solutions and the set of diverse solutions. The worst solution in *RefSet* is replaced with the best trial solution at any given iteration.

Subset Generation Method

This method generates subsets with two, three or more elements in a relatively reduced computational effort. To eliminate repetition of the elements in the subsets, the reference set with diverse solutions is used for the two by two combinations, instead of the complete reference set. The method also includes a new feature by adding a distant (or diverse) solution maximizing the distance from the region defined as the union of the vertices in the solution's subset. In this way, a new point "far from" the solution cluster is obtained at each iteration to maintain an appropriate diversity of solutions in the reference set.

Combination Method

This method uses each subset generated in the Subset Generation Method and combines its solutions, returning one or more trial solutions. Solution combinations are created using a filter vector applied to the union of solutions, called λ-filter. The λ-filters are used in this implementation as a form of structured combinations of solutions. Instead of drifting within the solution space defined by the reference set, the SS procedure searches each region extensively by applying different λ-filters. Each λ-filter generates a trial solution to be improved by the Improvement Method. A sequence of λ-filters applications generates a set of solutions within and beyond regions defined by two or more solutions in which new trial solutions will be chosen for updating the reference set in an evolutionary fashion.

Computational results obtained on a set of the most challenging clique DIMACS benchmark instances showed that the scatter search implementation compared well with some of the most competitive algorithms for the MCP.

5.   GRAPH COLORING

Graph k-coloring can be stated as follows: given an undirected graph G with a set V of vertices and a set E of edges connecting vertices, k-coloring G means finding a partition of V into k classes $V_1, ..., V_k$, called *color classes*, such that no couple of vertices $(u, v) \in E$ belongs to the same color class. Formally, $\{V_1, ..., V_k\}$ is a valid k-coloring of the graph $G = (V, E)$ if $\forall i \in [1..k]$ and $\forall (u, v) \in V_i, (u, v) \notin E$. The graph coloring problem (GCP) is the optimization problem associated with k-coloring. It aims at searching for the minimal k such that a proper k-coloring exists. This minimum is the chromatic number $\chi(G)$ of graph G.

Graph coloring has many real applications, e.g., timetable construction, frequency assignment, register allocation or printed circuit testing. Some of the solution methods for this problem include: greedy constructive approaches (DSATUR, RLF), hybrid strategies (HCA), simulated annealing, tabu search, GRASP or neural network. Some recent developments can be found in Laguna and Martí (2001a) and González-Velarde and Laguna (2001). The scatter search approach by Hamiez and Hao (2001) based on the template by Glover (1998) consist of the following elements.

- A **Diversification Generation Method** that uses *independent sets* to build initial configurations. Color classes are built one by one by selecting vertices in a random order to insure diversity.
- An **Improvement Method** based on the tabu search algorithm of Dorne and Hao (1998). This algorithm iteratively changes the current color of a conflicting vertex to another one, until achieving a proper coloring. A tabu move leading to a configuration better than the best configuration found so far, within the same execution of the Improvement Method or within the overall scatter search procedure, is always accepted as mandated by an *aspiration criterion*.
- A **Reference Set Update Method** that has been customized for this setting. Although this method is usually a "generic" element of scatter search, in this implementation the diversity measure takes on a special form. This point is crucial since, in the context of graph coloring, the Hamming distance is not well suited to compare two configurations c_1 and c_2. The distance between c_1 and c_2 is the minimum number of moves necessary to transform c_1 into c_2. The fitness of any configuration is the number of conflicting edges.
- A **Solution Combination Method** that uses a generalization of the powerful greedy partition crossover (GPX), proposed by Galinier and Hao (1999) within an evolutionary algorithm. GPX has been especially developed for the graph coloring problem with results reaching, and sometimes improving, those of the best known algorithms for the GCP. Given a subset s generated by the Subset Generation Method, the generalized combination operator builds the k color classes of the new configuration one by one. First, it chooses a configuration $c \in s$. Then, it removes from c a minimal set of conflicting vertices such that c becomes a partial proper k-coloring. Next, it fills in a free color class of the new configuration with all conflict-free vertices of the color class with maximum cardinality in c. It repeats these steps until the k color classes of the new configuration contain at least one vertex. Finally, to complete the new configuration if necessary, it assigns to each free vertex a color such that it minimizes the conflicts in the graph.

Computational testing was carried out on some of the well-known DIMACS benchmark graphs (Johnson and Trick, 1996). The scatter search procedure (SSGC) was compared with the generic tabu search (GTS) algorithm of Dorne and Hao (1998) together with the best-known methods available for the graph coloring problem due to Morgenstern (1996) and Funabiki and Higashino (2000).

The scatter search approach SSGC managed to reach the results of the best-known algorithms in quality (minimal number of colors used), except on the r1000.5 graph for which a 237-coloring was found by Funabiki and Higashino (2000). (The sophisticated algorithm used to reach this coloring includes, among other components, the search for a maximum clique.) Nevertheless, SSGC obtained a better coloring (240) than GTS (242) and outperformed the previous best result (241) for this graph (Morgenstern, 1996). This scatter search approach also improves in quality on the results obtained with tabu search (GTS) on a few other graphs.

6.   PERIODIC VEHICLE LOADING

Delgado, Laguna and Pacheco (2002) address a logistical problem of a manufacturer of auto parts in the north of Spain. The manufacturer stores auto parts in its warehouse until customers retrieve them. The customers and the manufacturer agree upon an order pickup frequency. The problem is to find the best pickup schedule, which consists of the days and times during the day that each customer is expected to retrieve his/her order. For a given planning horizon, the optimization problem is to minimize the labor requirements to load the vehicles that the customers use to pick up their orders.

Heuristically, the authors approach this situation as a decision problem in two levels. In the first level, customers are assigned to a calendar, consisting of a set of days with the required frequency during the planning horizon. Then, for each day, the decision at the second level is to assign each customer to a time slot. The busiest time slot determines the labor requirement for a given day. Therefore, once customers have been assigned to particular days in the planning horizon, the second-level decision problem is equivalent to a multiprocessor scheduling problem (MSP), where each time slot is the equivalent of a processor, and where the objective is to minimize the makespan.

A scatter search procedure is developed for the problem of minimizing labor requirements in this periodic vehicle-loading problem and artificial as well as real data are used to assess its performance. The scatter search constructs and combines calendar assignments and uses a heuristic to solve the MSP's for each day in the planning horizon and thus obtain a complete solution.

The Diversification Generation Method is based on GRASP constructions. The greedy function calculates the increase in labor requirements from assigning a previously unassigned order to a calendar. The procedure starts with all the orders in the unassigned set. The orders are considered one by one, from the largest to the smallest (i.e., from the one that requires the most amount of labor to the one that requires the least amount of labor).

The Improvement Method is based on a procedure that changes the assignment of an order from its current calendar to another. Preliminary experiments showed that the performance of the Improving Method with simple moves (i.e., the change of calendars for one order only) was not as good as the performance of a local search employing composite moves. A composite move is a chain of simple moves. Therefore, while a simple move prescribes the change of one order from one calendar to another, a composite move prescribes the change of several orders from their current calendars to others. It may seem that a local search based on simple moves should be capable of finding sequences of moves that are equivalent to composite moves. However, this is not necessarily the case because the local search based on simple moves is greedy and searches for the best exchange and performs the exchange only if it results in an improving move. A local search with composite moves, on the other hand, may perform some non-improving simple moves that lead to a large improving move.

The Combination Method generates new solutions by combining the calendar assignments of two reference solutions. The objective function values of the reference solutions being combined are used to probabilistically assign orders to calendars in the new trial solution. That is, on the average, most of the assignments come from the reference solution with the better objective function value. The procedure uses a static update of the reference set.

Using both randomly generated data adapted from the literature and real data from a manufacturer, the authors were able to show the merit of the scatter search design. In particular, extensive experiments showed that significant savings may be realized when replacing the manufacturer's current rules of thumb with the proposed procedure for planning purposes.

7.   CAPACITATED MULTICOMODITY NETWORK DESIGN

The fixed-charge capacitated multicommodity network design formulation (CMND) represents a generic model for a wide range of applications in planning the construction, development, improvement, and operations of transportation, logistics, telecommunication, and production systems, as well as in many other major areas. The problem is usually modeled as a combinatorial optimization problem and is NP-hard in the strong sense. Thus, not only the generation of optimal solutions to large problem instances constitutes a significant challenge, but even identifying efficiently good feasible solutions has proved a formidable task not entirely mastered.

The goal of a CMND formulation is to find the optimal configuration — the links to include in the final design— of a network of limited capacity to satisfy the demand of transportation of different commodities sharing the network. The objective is to minimize the total system cost, computed as the sum of the link fixed and routing costs.

Ghamlouche, Crainic and Gendreau (2001) proposed a new class of cycle-based neighborhood structures for the CMND and evaluated the approach within a very simple tabu-based local search procedure that currently appears as the best approximate solution method for the CMND in terms of robust performance, solution quality, and computing efficiency. Still more recently, Ghamlouche, Crainic and Gendreau (2002) explore the adaptation of path relinking to the CMND. This work evaluates the benefits of combining the cycle-based neighborhood structures and the path relinking framework into a better meta-heuristic for this difficult problem.

The method proceeds with a sequence of cycle-based tabu search phases that investigate each visited solution and add elite ones to *RefSet*. When a predefined number of consecutive moves without improvement is observed, the method switches to a path relinking phase. Six different strategies to construct *RefSet* are considered:

- In strategy **S1**, *RefSet* is built using each solution that, at some stage of the tabu search phase, improves the best overall solution and becomes the best one.
- Strategy **S2** retains the best local minimum found during the tabu search phase. This strategy is motivated by the idea that local minimal solutions share characteristics with global optimal solutions.
- Strategy **S3** selects *R-improving* local minima, that is, local minimal solutions that have a better evaluation of the objective function than those already in *RefSet*.
- Strategy **S4** allows solutions to be retained in *RefSet* not only according to an attractive solution value but also according to a diversity or dissimilarity criterion.
- Strategy **S5** aims to ensure both the quality and the diversity of solutions in *RefSet*. Starting with a large set of "good" solutions, *RefSet* is partially filled with the best solutions found, to satisfy the purpose of quality. It is then extended with solutions that change significantly the structure of the solutions already in *RefSet* to ensure diversity.
- Strategy **S6** proceeds similarly to S5 with the difference that *RefSet* is extended with solutions close to those already in it.

During the path relinking phase, moves from the initiating solution to a neighbor one direct the search towards the guiding solution. Due to the nature of the neighborhoods used, there is no guarantee that the guiding solution will be reached. One cannot, therefore, stop the process only if the current and the guiding solutions are the same. Hence, the procedure computes Δ_{IG} as the number of arcs with different status between the initiating and the guiding solutions and allows the search to explore a number of solutions not larger than Δ_{IG}.

Initiating and guiding solutions are chosen from *RefSet*. The criteria to select them are also critical to the quality of the new solutions and, thus, the performance of the procedure. The authors investigated the effect of the following criteria:

C1: Guiding and initiating solutions are defined as the best and worst solutions, respectively.

C2: Guiding solution is defined as the best solution in the reference set, while the initiating solution is the second best one.

C3: Guiding solution is defined as the best solution in the reference set, while the initiating solution is defined as the solution with maximum Hamming distance from the guiding solution.

C4: Guiding and initiating solutions are chosen randomly from the reference set.

C5: Guiding and initiating solutions are chosen as the most distant solutions in the reference set.

C6: Guiding and initiating solutions are defined respectively as the worst and the best solutions in the reference set.

The path relinking phase stops when the reference set becomes empty (cardinality ≤ 1). Then, either stopping conditions are verified, or the procedure is repeated to build a new reference set.

Extensive computational experiments, conducted on one of the 400 MHz processors of a Sun Enterprise 10000, indicate that the path relinking procedure offers excellent results. It systematically outperforms the cycle-based tabu search method in both solution quality and computational effort. On average, for 159 problems path relinking obtains a gap of 2.91% from the best solutions found by branch-and-bound versus a gap of 3.69% for the cycle-based tabu search. The branch and bound code, CPLEX 6.5, was allowed to run for 10 CPU hours. Thus, path relinking offers the best current meta-heuristic for the CMND.

8.   JOB-SHOP SCHEDULING

Job-shop scheduling is known to be a particularly hard combinatorial optimization problem. It arises from operations research practice, has a relatively simple formulation, excellent industrial applications, a finite but potentially astronomical number of solutions and unfortunately is NP-hard in the strong sense. It is also considered a good benchmark to measure the practical efficiency of advanced scheduling algorithms. In the early nineties, after a series of works dealing with optimization algorithms of the branch and bound (B&B) type, it became clear that pure optimization methods for this problem had a ceiling on their performance. In spite of important advances over the past two decades, the best B&B methods cannot solve instances with more than 200 operations in a reasonable time (hours, days, weeks).

A new era started when job-shop algorithms based on the TS approach appeared. The simple and ascetic algorithm TSAB (Nowicki and Smutnicki, 1996), designed originally in 1993, found the optimal solution of the notorious job-shop instance FT10 (100 operations) in a few seconds on a PC. This instance had waited 26 years, since 1963, to be solved by an optimization algorithm. But going far beyond the solution of FT10, the TSAB approach made it possible to solve, in a very short time on a PC, instances of size up to 2,000 operations with unprecedented accuracy — producing a deviation from an optimality bound of less than 4% on average. This is considerably better than the deviation of approximately 20% for special insertion techniques, 35% for standard priority rules and over 130% for random solutions. Another highly effective tabu search method for the job shop problem was introduced by Grabowski and Wodecki (2001).

Further exploration of the ideas underlying TSAB focuses on two independent subjects: (1) acceleration of the speed of the algorithm or some its components, and (2) a more sophisticated diversification mechanism, the key for advanced search scattering. Additional papers by Nowicki and Smutnicki (2001a and 2001b) provide some original proposals that follow these research streams. They refer to a new look at the landscape and valleys in the solution space, set against the background of theoretical properties of various distance measures. There are proposed accelerators based on theoretical properties, which, by means of skillful decomposition and aggregation of calculations, significantly speed up the search process. These accelerator are: (a) INSA accelerator (advanced implementation of insertion algorithm used for starting solutions in TSAB), (b) tabu status accelerator, (c) NSP accelerator (fast single neighborhood search). Next, in order to diversify the search, TSAB has been embedded in the scatter search and path relinking frameworks. The resulting algorithm i-TSAB described in Nowicki and Smutnicki (2001a), the powerful successor of TSAB, works with elite centers of local search areas forming a milestone structure, modified by space explorations conducted from viewpoints located on gops (a class of goal oriented paths).

As the immediate practical result of this new approach, better upper bounds (new best solutions) have been found for 24 of the 35 instances from the common benchmark set of Taillard, attacked by all job-shop algorithms designed till now. The proposed algorithm still runs on a standard PC in a time of minutes.

9. CAPACITATED CHINESE POSTMAN PROBLEM

The problem treated in Greistorfer (2001a) is the so-called *capacitated Chinese postman problem* (CCPP). The goal of the (undirected) CCPP is to determine a least-cost schedule of routes in an undirected network under the restriction of a given fleet of vehicles with identical capacity, which operates from a single depot node. In the standard version of the CCPP the number of vehicles is unlimited, i.e. determining the number of vehicles is a decision variable. The CCPP is a special instance of the general class of arc routing problems, a group of routing problems where the demand is located on arcs or edges (one-way or two-way roads) connecting a pair of nodes (junctions). Relevant practical examples of the CCPP are postal mail delivery, school bus routing, road cleaning, winter gritting or household refuse collection. But applications are not limited to the routing of creatures or goods. There are also cases in industrial manufacturing, e. g. the routing of automatic machines that put conducting layers or components on to a printed circuit board.

The algorithmic backbone of a *tabu scatter search* (TSS) metaheuristic developed for the solution of the CCPP is a tabu search (TS) that employs a set of neighborhood operations, defined by edge exchanges and insert moves, and a long-term diversification strategy guided by frequency counts. The short-term memory mechanism is based on attributes defined by edges that simply prohibit reversal of moves for a dynamically determined tenure. Additionally, the procedure has a solution pool component that maintains a set of elite solutions found in the course of the optimization. If classic genetic algorithms are classified as pure parallel pool methods because they work with a set of high quality solutions, then the TSS follows a sequential pool design, where periods of isolated and single-solution improvements of TS alternate with multi-solution combinations. The type of encoding used turns the Combination Method into one that is typical to pure scatter search implementations. The TSS architecture as proposed by Greistorfer (2001a) does not exactly follow the ideas of Glover's (1998) template, although there are many common features as outlined below.

The Combination Method combines elite solutions which have been collected during the normal operation of the tabu search. As suggested in the template paper and subsequent publications, the combination of solutions is supported by generalized rounding procedures. The underlying principle of the Combination Method developed in this context is an adaptation of the formulation of a transportation problem that generalizes the assignment operator of Cung, et al. (1997).

The Combination Method benefits from a data structure that is typically used in problems defined by pure permutations (customers) or by permutations where sub-strings (routes) have to be considered as well. The Combination Method works as follows. Given a set of elite solutions $s_1,...,s_c$, the coefficients a_{ij} for the transportation problem formulation denote the number of times customer j is assigned to route i in the subset under consideration. The coefficient matrix associated with the transportation problem may be interpreted as an assignment frequency matrix that is the linear combination of c individual assignment matrices. Unit demands in the transportation problem formulation reflect the need to serve every edge in the network with a single vehicle. Route supplies are approximated with the average number of customers that can be serviced with exceeding the vehicle capacity. A dummy column is added to pick up the oversupply. The solution of the resulting transportation problem yields customer-route assignments that maximize the total number of desirable assignments while simultaneously minimizing the total Euclidean distance in the assignment frequency matrix. The outcome of this Combination Method is subjected to a greedy sequencing heuristic that finds a local optimal assignment and sequence of customers in all routes.

The overall procedure starts from a random set of solutions, which are gradually exchanged for solutions found during the execution of the TS phase. The Combination method is occasionally called to construct a trial solution that triggers a TS phase. This process that alternates between TS and the Combination Method stops after a pre-defined period of iterations. The TSS was tested on several classes of CCPP instances:

– planar Euclidean grid-graph instances
– Euclidean random instances
– the well-known DeArmon data set

In a direct comparison with an old TS method (see Greistorfer (1995)) known as CARPET, TSS significantly improves the results for the Euclidean problems (in 54% of all cases) and is competitive regarding the instances from literature. In particular, TSS is able to find all (known) optimal solutions and establish one more best-known solution. Its worst average deviation (due to a single instance) is only 1.29% higher than the one of CARPET. The total running times, which are scaled with respect to CARPET-PC, are longer. However, it is shown that, on average, TSS obtains its best results faster than the CARPET heuristic. Thus, adding a pool component to a TS and using an advanced Combination Method resulted in an improved solution procedure.

9.1   Testing population designs

The research focus in Greistorfer (2001a) is continued in Greistorfer (2001b), which emphasizes advances in methodology. The main research task is to design strategies to manipulate the pool of solutions and to evaluate them by means of thorough computational comparisons. Test results once again refer to a sample of CCPP arc routing instances but, as mentioned above, the encoding makes it possible the generalization to other problem settings. From the variety of design options associated with pool methods, our discussion concentrates on three basic components: the input function (equivalent to the Reference Set Update Method in SS) and output function (equivalent to the Subset Generation Method in SS), which are responsible for pool maintenance and which determine the transfer of elite solutions, and a solution Combination Method that must effectively combine a set of elite solutions provided by the output function.

The variants of the TSS are comprised of four input strategies, $I_{0,...,3}$, four output strategies, $O_{0,...,3}$, and three Combination Methods, denoted by $M_{0,...,2}$. The variant indexed 0 corresponds to the settings proposed by Greistorfer (2001a).

Input strategy I_0 limits the definition of solution quality to the value of the objective function and ignores the structural properties of solutions and their interrelationships. I_1 overcomes this disadvantage by including full duplication checks between potential elite solutions and pool members. I_2 resembles the Reference Set Update method of Glover (1998) and partially uses hashing for duplication checking. In I_3 an attempt is made of finding a compromise that skips the full duplication of I_2 and relies solely on hashing.

Similarly to I_0, O_0 does not utilize structural information and simply relies on random selection. O_1 uses frequency counts and selects those solutions for combination which have not been picked before or have been rarely picked. This frequency memory introduces an effect that contrasts with selection mechanisms, such as those employed in genetic algorithms, where memory is an implicitly function. By contrast, the Subset Generation Method in scatter search explicitly avoids the generation of solution subsets that have already been examined in the past. Output strategies O_2 and O_3 select solutions which have the smallest and largest distance to each other, respectively. The distance measure is based on aggregating the positions of the customers and their route membership.

The combination strategy M_0 is the LP-based transportation method described in detail in Greistorfer (2001a). The other two Combination Methods are based on constructing average solutions. M_1 constructs average customer labels whereas M_2 determines average customer positions (see also Campos et al. (1999)). Both approaches relax the capacity restriction, which is then later enforced by splitting the permutation sequence into a set of route clusters. The trial solution is obtained after applying the post-optimizing greedy sequencing heuristic.

The computational investigation of the results for the different TSS designs was performed by comparing the performance of all possible 4·4·3=48 configurations against each other. Each configuration was tested on the entire set of instances and evaluated by the average objective function value of the best solutions found. After executing all these experiments, the best configuration turned out to be (I_3, O_3, M_0). In order to evaluate the specific effects of an input, output or combination variant, an analysis of variance was performed with SPSS.

Generally, effects of variations tended to be smaller at the input side of the pool since all tests did not indicate any significant difference among the tested input strategies. One possible explanation is that input procedures $I_{1,2,3}$ effectively prevent duplications. Another possible explanation might be that input strategies that encourage a high level of diversification are not adequately utilized by the other search components. Finally, the results generated by the straightforward input function I_0 are generally inferior to the results obtained when using any of the other input functions.

Regarding output functions and Combination methods the findings are mixed. It was found that the min-distance approach in O_2 is definitely an inferior option. The argument that good solutions are likely to be found in the vicinity of the best solutions is not supported by the results of these experiments. The SS philosophy of selecting diverse solutions to be combined is therefore particularly useful. The superior performance of the max-distance functions O_3 over O_2 supports this finding. The expected effect of the use of memory in O_1 was not significant. Random sampling in O_0 can be justified when viewed in isolation by ignoring the effect of the input function and combination method.

Finally, The LP approach of M_0 significantly contributes to finding better solutions than M_1 and M_2. While there are no significant relations between M_1 and M_2, the individual best choice for M_0 plays a useful role in the collective optimal design (I_3, O_3, M_0).

10.   VEHICLE ROUTING

The classical Vehicle Routing Problem (VRP) can be defined as follows. Let $G=(V,A)$ be a graph where $V=\{v_0, v_1, ..., v_n\}$ is the vertex set, and $A=\{(v_i,v_j) \mid v_i,v_j \in V, i \neq j\}$ is the arc set. Vertex v_0 denotes a depot, where a fleet of m identical vehicles of capacity Q are based, and the remaining vertices represent n cities (or client locations). With each vertex v_i $(i=1,..,n)$ is associated a quantity q_i of some goods to be delivered by a vehicle. A nonnegative distance or cost matrix $C=\{c_{ij}\}$ is defined on A. The VRP consists of determining a set of m vehicle routes of minimal total cost, starting and ending at a depot v_0, such that every vertex v_i is visited exactly once by one vehicle and the total quantity assigned to each route does not exceed the capacity Q of the vehicle which services the route. A solution to the VRP is defined as a set of m routes, $S=\{R_1, R_2, ..R_m\}$ where R_k is an ordered set representing consecutive vertices in the route k.

Rego and Leao (2000) propose a scatter search implementation for the VRP problem. This tutorial paper does not try to improve upon the best published methods for the VRP but it provides an illustration of how to apply the scatter search methodology to a hard combinatorial problem.

The approach constructs solutions from permutations that are generated with the method described in Glover (1997). A solution is derived from a permutation by first constructing clusters, where each cluster of vertices (a route) is obtained by successively assigning a vertex v_i to a route R_h as long as the sum of the corresponding q_i values does not exceed Q. As soon as such a cutoff limit is attain a new assignment (route) is created. A straightforward method to create this solution consists of successively linking vertices in the order they appear in a permutation and attaching the initial and ending vertices to the depot. Then, a 2-opt procedure with a first improvement strategy is applied to improve each route. The set P contains all the different solutions generated.

A simple rule is applied to create and update *RefSet*, where intensification is achieved by the selection of high-quality solutions (in terms of the objective function value) and diversification is induced by including diverse solutions from the current set of new trial solutions. Thus, the reference set can be defined by two distinct subsets *RefSet_1* and *RefSet_2*, representing respectively the subsets of high-quality and diverse solutions. The method starts by including in *RefSet* the best b_1 solutions in P. In this context the terms "highest evaluated solution" and "best solution" are interchangeable and refer to the solution that best fits the evaluation criterion under consideration. Then, candidate solutions are included in *RefSet* according to the max-min criterion described in the tutorial chapters.

The Subset Generation Method considers the four subsets types described in Section 2 of Chapter 5. New solutions are generated by weighted linear combinations associated with the subsets under consideration. One solution is generated for each subset by a convex linear combination based on scores and then rounding of the variables. A new solution is created using the edges associated with those variables that have a value of 1. The set of these edges does not usually represent a feasible graph structure for a VRP solution. That is, it might produce a subgraph containing vertices with a degree different than two. Such subgraphs can be viewed as fragments of solutions (or partial routes). To create a feasible solution, vertices which have a degree equal to 1 are directly linked to the depot. This forces to maintain subgraph feasibility throughout the search. It is also possible that the subgraph resulting from a linear combination contains vertices of degree greater than two. In this case, a straightforward procedure is applied that consists of successively dropping edges with the smallest scores in the star of these vertices until their degree becomes equal to two. By doing so, the subgraph obtained will be either feasible or fall into the case where some of the vertices have degree 1, which can be handled as already indicated. Even though the Combination Method creates feasible subgraphs for the VRP, the solution may still not be feasible in relation to the other problem constraints. Therefore, the Improvement Method must be able to deal with infeasible solutions.

The Improvement Method works in two stages. The first stage is concerned with making the solution feasible while choosing the most favorable move (relative to the objective function cost), and the second stage is the improvement 2-opt process that operates only on feasible solutions. In general, several variants can be used to deal with infeasible solutions. These techniques are usually based on varying penalty factors associated with the problem constraints. Some constraints are potentially "less damaging" when violated than others, and usually the choice of penalty values should take this into consideration. Also, the way these penalties are modified can make the search more or less aggressive for moving into the feasible region. High penalty values are usually employed for an intensification strategy, and lower values for a diversification approach that allows the search to remain longer in the infeasible region.

Rego and Leao (2000) consider a method that doesn't use a penalty factor but rather identifies the most violated route and makes the best (cost reducing) move that consists of removing a vertex from this route and feasibly inserting it in another route. It is possible that an additional route will be created if such a move is not feasible or if all routes are over their capacity.

These scatter search variants for the solution of the VRP are illustrated on an example with 14 locations and a number of vehicles ranging from 1 to 14. Several strategies and implementations are suggested for each method.

11.   BINARY MIXED INTEGER PROGRAMMING

Linear programming models with a mixture of real-valued and binary variables are often appropriate in strategic planning, production planning with significant setup times, personnel scheduling and a host of other applications. The abstract formulation for linear problems with binary integers takes as data a row vector c, of length n, a $m \times n$ matrix A and a column vector b of length m. Let D be the index set $1, ..., n$. The problem is to select a column vector x of length n so as to:

$$\min \sum_{i \in I} c_i x_i$$

s.t.

$$Ax \geq b$$
$$x_i \in \{0,1\} \qquad i \in I$$
$$x_i \geq 0 \qquad i \in D \setminus I$$

where the index set I consists of the variables that must take on zero-one values.

11.1   Pivot Based Search with Branch and Bound

Issues related to the behavior of a pivot based tabu search integrated with branch and bound algorithm, using path relinking and chunking are discussed by Løkketangen and Woodruff (2000). The integration takes place primarily in the form of local searches launched from the nodes of the branch and bound (B&B) tree. These searches are terminated when an integer feasible solution is found or after some number of pivots, NI. Any time a new best solution is found, the search is continued for an additional NI pivots. Chunking (see Woodruff 1996, 1998) is used to detect solutions that should be used for special path relinking searches that begin at the LP relaxation and to determine when the use of pivot searches should be discontinued. (See also Glover, Løkketangen and Woodruff, 2000, for another application of chunking to the same kind of problems.)

As the search is launched from nodes in a B&B tree, there are some special considerations that come into play that sets this use of the pivot based search somewhat apart from other implementations. First, the chunking mechanism and the path relinking based target searches, respectively, fulfill the functions of diversification and intensification. Second, the purpose, or focus, of the search is somewhat different from the stand-alone search, in that for some of the searches, the emphasis is shifted more towards obtaining integer feasibility quickly. This focus is controlled by a separate parameter, *skew*, that is used to adjust the relative importance of obtaining feasibility versus maintaining a good objective function value.

Chunking addresses the questions of when the launching of pivot based searches should be terminated, and when the path relinking searches should be launched. More specifically, path relinking searches are used to exploit "unique" or "outlying" solutions. The meaning of "unique" and "outlying" can be supplied by chunking.

Two types of local searches can be launched at a node. The first are the normal TS pivot-based searches launched from nodes in the B&B tree (see Løkketangen and Glover, 1995, 1996, 1998, 1999). The other types are the path relinking searches. After a best-so-far solution x^* has been found, the chunking mechanisms try to identify distant solutions, x', with respect to the current sample. When such a distant solution has been identified, a *2-target* search is launched. This is a variant of path relinking, with the purpose of launching a new search into unknown territory, while at the same time keeping good parts of the solutions. This integrates the considerations of intensification and diversification.

The starting point of this search is the relaxed root node solution LP^* (being an upper bound on the objective function value), and the target for the path relinking is the hyperplane defined by the common integer solution values of x^* and x'. All integer variables are freed. To allow the search to focus on this hyperplane, the integer infeasibility part of the move evaluation function is temporarily modified. (The objective function value component is unaltered, as is the aspiration criterion —see Løkketangen and Woodruff, 2000.) Instead of using the normal integer infeasibility measure of summing up over all the integer variables the distance to the nearest integer, the authors use the following scheme:

- Sum up over all the integer variables.
- If the two targets have the same integer solution value for the variable, use the distance to this value.
- If the two targets differ, use the normal integer infeasibility measure (i.e. the closest integer value).

When the search reaches the hyperplane connecting x^* and x', the normal move evaluation function is reinstated, and the search continues in normal fashion for *NI* iterations.

Computational testing was done on problems from Miplib and Dash Associates, consisting of a mix of MIP's and IP's. The testing showed that the local searches had a beneficial effect on the overall search time for a number of problem instances, particularly those that have historically been harder to solve.

11.2   Generating Diverse Solutions

Often, scatter search and star path algorithms (Glover 1995), generate diverse sets of solutions as a means to an end. In a recent paper by Glover, Løkketangen and Woodruff (2000) diversity is the ultimate goal for which scatter search and star paths are employed. This paper presents methods of systematically uncovering a diverse set of solutions for 0-1 mixed integer programming problems. These methods can be applied to instances without any special foreknowledge concerning the characteristics of the instances, but the absence of such knowledge gives rise to a need for general methods to assess diversity.

When the objective function is only an approximation of the actual goals of the organization and its stakeholders, the one solution that optimizes it may be no more interesting than other solutions that provide good values. However, information overload can be a problem here as well. It is not desirable to swamp the decision maker with solutions. Highly preferable is to identify a set of solutions that are decently good and, especially, diverse. One can reasonably rely on the objective function to quantify the notion of "decently good". The diversification methods given by Glover, Løkketangen and Woodruff (2000) are based on the idea of generating extreme points in a polyhedral region of interest and then using these points and the paths between them in a variety of ways. The methods examine points on the polyhedron, within and "near" it. Their algorithm proceeds in two phases: first it generates a set of *centers* and then connects them using *star paths*.

The description of the generation of centers can also be broken into two phases. First a diversification generator is used to create points. In the second phase, these points are provided as data to an optimization problem that results in extreme points that are averaged to create the centers.

Although a diverse set of good solutions is clearly desirable, it is not clear in advance how to measure the property of diversity. In spite of the fact that the objective function is not exact, it presumably gives a reasonable way to assess the relative "goodness" of a set of solutions. No such simple mapping is known from solution vectors to a one-dimensional measure of diversity. Diversity measures are required both for the design of practical software and for research purposes. For practical software, it is important to know if the user should be "bothered" with a particular solution vector — that is, to know if a vector adds enough diversity to warrant adding it to the set of solutions that are displayed. For research purposes, one might want to compare the set of vectors generated by one method with a set of vectors generated by another.

There are a number of advantages to the quadratic metric known in this context as *Mahalanobis* distances. This metric is scale invariant and can take correlations into account if based on a covariance matrix. Furthermore, this type of distance connects naturally with a scalar measure of the diversity of a set of vectors, which is the determinant of the covariance matrix of the set. Under the assumption of multivariate normality, the covariance matrix defines ellipsoids of constant Mahalanobis distances that constitute probability contours. Large covariance determinants correspond to large volumes in the ellipsoids. The assumption of multivariate normality is not needed to use the covariance determinant to put an order on sets of vectors and furthermore it is not needed to see that adding points with large Mahalanobis distances will increase the covariance determinant.

However, there is a major difficulty. In order to calculate a covariance matrix for a set of vectors of length $p = n$ one must have a set of vectors that does not lie entirely in a subspace. This means that at a minimum the set must contain $n + 1$ vectors and for MIP solutions, more vectors will often be required to span the full n dimensions. For even modest sized MIPs this is impractical. In order to have a working definition of diversity, one must have thousands of solution vectors. A remedy for this difficulty that also increases the plausibility of multivariate normality has been referred to as *chunking* by Woodruff (1998). A generalization based on principal components has also been proposed by Woodruff (2001).

As general purpose optimization methods are embedded in decision support systems, there will unquestionably be an increased need not only for optimal solutions, but also for a diverse set of good solutions. Scatter search and star paths can be an effective means to this end.

Results of computational experiments demonstrate the efficacy of the *scatter-star-path* method for generating good, diverse vectors for MIP problems. Furthermore, the results show that the method offers particular advantages when used in conjunction with branch and bound. The creation of these results illustrates the use of methods for measuring the diversity for a set of solutions.

12.   ITERATED RE-START PROCEDURES

Research has been performed to investigate the ability of path relinking to improve the performance of iterated re-start procedures, with attention focused in particular on GRASP (Resende and Ribeiro, 2002). One possible shortcoming of standard GRASP is the independence of its iterations, i.e., the fact that it does not learn from the history of solutions found in previous iterations. This is so because it discards information about any solution encountered that does not improve the incumbent. Information gathered from good solutions can be used to implement extensions based on path-relinking.

Path relinking was originally proposed in the context of tabu search as an intensification strategy which explores trajectories connecting high-quality solutions. The use of path relinking within GRASP was first proposed by Laguna and Martí (1999), being followed by several extensions, improvements, and successful applications (e.g., Canuto, et al. 2001; Ribeiro, et al. 2002). Path relinking and a very short term memory used within the local search were instrumental to make a recently proposed GRASP heuristic for the capacitated minimum spanning tree problem competitive with other approaches in the literature (Souza, et al. 2002). Two basic strategies are used to apply path relinking in the context of a GRASP heuristic:

- applying path relinking as a post-optimization step to all pairs of elite solutions; and
- applying path relinking as an intensification strategy to each local optimum obtained after the local search phase

Both strategies maintain and handle a pool with a limited number *MaxElite* of elite solutions found along the search (*MaxElite* is set to values ranging from 10 to 20 in most implementations). The pool is originally empty. Each locally optimal solution obtained by local search is considered as a candidate to be inserted into the pool if it is sufficiently different from every other solution currently in the pool. If the pool already has *MaxElite* solutions and the candidate is better than the worst of them, then the former replaces the latter. If the pool is not full, the candidate is simply inserted.

Applying path relinking as an intensification strategy to each local optimum seems to be more effective than simply using it as a post-optimization step. In this context, path relinking is applied to pairs $x - y$ of solutions, where x is the locally optimal solution obtained after the application of a local search and y is one of a few elite solutions randomly chosen from the pool (usually only one elite solution is selected). The algorithm starts by computing the symmetric difference between x and y, resulting in a set Δ of moves which should be applied to one of them (the initial solution) to reach the other (the guiding solution). Starting from the initial solution, the best move still in Δ is applied, until the guiding solution is attained. The best solution found along this trajectory is also considered as a candidate for insertion in the pool and the incumbent is updated. Several alternatives have been considered and combined in recent implementations to explore trajectories connecting x and y:

- do not apply path relinking at every GRASP iteration, but instead only periodically;
- explore two different trajectories, using first x, then y as the initiating solution;
- explore only one trajectory, starting from either x or y; and
- do not follow the full trajectory, but instead only part of it.

All these alternatives involve trade-offs between computation time and solution quality. Ribeiro, et al. (2002) observed that exploring two different trajectories for each pair $x - y$ takes approximately twice the time needed to explore only one of them, with very marginal improvements in solution quality. They also observed that if only one trajectory is to be investigated, better solutions are found when path relinking starts from the best between x and y. Since the neighborhood of the initiating solution is much more carefully explored than that of the guiding one, starting from the best one gives the algorithm a better chance to investigate more fully the neighborhood of the most promising solution. For the same reason, the best solutions are usually found closer to the initiating solution than to the guiding one, allowing pruning the relinking trajectory before the latter is reached. The same findings were also observed on a recent implementation of a GRASP heuristic for a multicommodity flow problem arising from PVC rerouting in frame relay services. Detailed computational results and implementation strategies are described by Resende and Ribeiro (2002).

Path relinking is a quite effective strategy to introduce memory in GRASP, leading to very robust implementations. This is illustrated by the results obtained with the hybrid GRASP with path relinking for the Steiner problem in graphs described in Ribeiro, et al. (2002). This implementation was capable of improving upon 33 out of the 41 still open problems in the i640 series of the SteinLib repository (Voss, et al. 2001).

Even though parallelism is not yet systematically used to speed up or to improve the effectiveness of metaheuristics, parallel implementations are very robust and abound in the literature (see e.g. Cung et al. (2001) for a recent survey). Most parallel implementations of GRASP follow the independent-thread multiple-walk strategy, based on the distribution of the iterations over the processors.

The efficiency of multiple-walk independent-thread parallel implementations of metaheuristics, running multiple copies of the same sequential algorithm, has been addressed by several authors. A given target value τ for the objective function is broadcasted to all processors which independently run the sequential algorithm. All processors halt immediately after one of them finds a solution with value at least as good as τ. The speedup is given by the ratio between the times needed to find a solution with value at least as good as τ, using respectively the sequential algorithm and the parallel implementation with ρ processors. These speedups are linear for a number of metaheuristics, including simulated annealing, iterated local search, and tabu search. This observation can be explained if the random variable *time to find a solution within some target value* is exponentially distributed (Verhoeven and Aarts, 1995). In this case, the probability of finding a solution within a given target value in time pt with a sequential algorithm is equal to that of finding a solution at least as good as the former in time t using ρ independent parallel processors, leading to linear speedups. An analogous proposition can be stated for a two parameter (shifted) exponential distribution.

Aiex, et al. (2002) have shown experimentally, by fitting a two-parameter exponential distribution, that the solution times for GRASP also have this property. This result was based on computational experiments involving GRASP implementations applied to 2400 instances of five different problems: maximum independent set, quadratic assignment, graph planarization, maximum weighted satisfiability, and maximum covering. The same result still holds when GRASP is implemented in conjunction with a post-optimization path relinking procedure.

In the case of *multiple-walk cooperative-thread* parallel strategies, the threads running in parallel exchange and share information collected along the trajectories they investigate. One expects not only to speed up the convergence to the best solution but, also, to find better solutions than independent-thread strategies. Cooperative-thread strategies may be implemented using path-relinking, combining elite solutions stored in a central pool with the local optima found by each processor at the end of each GRASP iteration. Canuto, et al. (2001) used path relinking to implement a parallel GRASP for the prize-collecting Steiner tree problem. A similar approach was recently adopted by Aiex, et al. (2000) for the 3-index assignment problem. Each processor, upon completing its iterations, applies path relinking to pairs of elite solutions stored in a pool, and each processor keeps its own local pool of elite solutions. The strategy used in Canuto (2000) is truly cooperative, since pairs of elite solutions from a centralized unique central pool are distributed to the processors which perform path relinking in parallel. Computational results obtained with implementations using MPI and running on a cluster of 32 Pentium II-400 processors and on a SGI Challenge computer with 28 196-MHz MIPS R10000 processor (Aiex, et al. 2000) show linear speedups and further illustrate the effectiveness of path relinking procedures used in conjunction with GRASP to improve the quality of the solutions found.

13. PARALLELIZATION FOR THE P-MEDIAN

García-López, et al. (2002a) propose the parallelization of SS to achieve either an increase of efficiency or an increase of exploration. In any parallelization some steps of the algorithm are distributed among the available processors. The increase of the efficiency is obtained by performing the same steps in less time than the sequential algorithm, whereas the increase of the exploration is obtained by performing more steps in the same time as the sequential algorithm. Several strategies for parallelizing a SS algorithm are analyzed. The methods are tested with large instances of the p-median problem obtained from the TSPLIB.

Given the set $L=\{v_1, v_2, ..., v_m\}$ of potential locations for the facilities (or location points), and the set $U=\{u_1, u_2, ..., u_n\}$ of users (or customers, or demand points), the entries of an $n \times m$ matrix $D=(d_{ij})_{n \times m}=(Dist(u_i, v_j))_{n \times m}$ give the distances traveled (or costs incurred) for satisfying the demand of the user located at u_i from the facility located at v_j, for all $v_j \in L$ and $u_i \in U$, the objective of the p-median problem is to minimize the sum of these distances (or transportation costs), i.e.,

$$\text{minimize} \sum_{u_i \in U} \min_{v_j \in X} Dist(u_i, v_j)$$

where $X \subseteq L$ and $|X| = p$.

The initial population is based on the scatter search principle of generating a set of disperse and good solutions. It consists of the following phases. The first one starts by dividing the set L in several sets. A constructive method to get a good solution consists of, from an arbitrary initial point u of L, select $p-1$ times the farthest point to the already selected points. This constructive method is applied for each set of the partition of L. Then a solution is obtained for each set of the partition. However, since the constructed solution depends on the starting point, it is applied from several starting points to obtain different solutions that are in turn submitted to the

Improvement Method. The dispersion among solutions is evaluated with a distance measure based on the objective function.

The reference set consists of a set of good and disperse solutions selected from the population. The generation of a reference set is done by selecting b_1 solutions from the best solutions and b_2 disperse solutions ($b=b_1+b_2$) as described in the tutorial chapters.

The Subset Generation Method is limited to the generation of every new solution pair. Given the subset of selected solutions from the reference set, the Combination Method tries to construct a new trial solution with the good characteristics of the reference solutions. Let X be the set of points that are in all these solutions. For every user point u let $L(u) = \{v \in L : Dist(u, v) \leq \beta\,Distmax\}$, where

$$Dist\max = \max_{u,v \in L} Dist(u,v).$$

Choose the point u^* in L such that $Dist(X, u^*) = max_{u \in L} Dist(X, u)$ and select at random a point $v \in L(u^*)$ that is included in X. This step is iteratively applied until X has size p.

The Improvement Method is based on swapping moves. Given a solution, the move consists of dropping an element from it, and adding a new one. The method performs a local search until no further improvement is possible.

Three different parallelization strategies for the SS implementation are considered. The first is a synchronous algorithm that enables solving, in parallel, the local searches; thus reducing the running time. The second parallelization is achieved by selecting several subsets from the reference set that are combined and improved by the processors. These procedures are replicated as many times as the number of available processors. The local optima found by the processors are used to update the reference. The last parallelization increases the exploration in the solution space by parallel running the sequential scatter search for several populations. It consists of a multi-start search where the local searches are replaced by SS using different populations that run on the parallel processors

The parallel algorithms were coded in C using OpenMP, (a model for parallel programming portable across shared memory architectures), and tested with large instances of the p-median problem. The distance matrix was taken from the instance TSPLIB RL1400 that includes 1400 points. The objective values found with these algorithms are comparable with the best obtained in the literature (García-López, et al. 2002b; Hansen, et al. 2001).

14. OPTQUEST APPLICATION

Optimization and simulation models play an important role in analysis of models for military defence. A crucial component of a model-based analysis effort is agreement upon the planning scenario upon which the analysis is conducted. For example in military force structuring, the model might suggest a prescribed force structure necessary to best meet the demands of a planning scenario. Conversely, given some proposed force structure, a model might provide insight into a "best use" of that force within the specified scenario. A particularly perplexing challenge for military analysts occurs when they must suggest a single overall force structure after considering multiple competing scenarios, each suggestive of potentially differing optimal force structures.

Hill and McIntyre (2000) addressed this particular vexing military force structure problem. They define a *robust force structure* (solution) as that force structure "that provides the best overall outcome as evaluated with respect to some set of scenarios each of which has an associated likelihood of occurrence." Their approach considered the multi-scenario optimization problem within which each particular scenario solution becomes a component of an aggregate multi-scenario solution. They treat the multi-scenario space as a composite of the component scenario spaces where each component space contributes relative to its likelihood of occurring or relative importance weight. Employing a metaheuristic to guide a search that executes a combat model to evaluate the objective function provides a means to find a single solution, potentially optimal in the multi-scenario space, and by definition, robust across each of the individual scenario spaces.

Central to Hill and McIntyre approach is a CONTROLLER interface between the metaheuristic module and the combat models conducting the evaluations. The metaheuristic guides the search process providing the CONTROLLER potential solutions (input force structure) and receiving from the CONTROLLER evaluations of those solutions. The COMBAT MODEL receives its input (and scenario) from the CONTROLLER, evaluates the input, and returns the required quality measure from the combat model assessment associated with the input. The CONTROLLER accepts the potential solutions, provides them to each of the scenario evaluators in the COMBAT MODEL, and combines each measure into the final value or fitness of the potential solution. This process continues until predefined stopping conditions are satisfied at which time the best, or set of best, solutions are returned.

Bulut (2001) applied scatter search, implemented within the OptQuest callable library, to solve a multi-scenario optimization problem based on the United States Air Force's Combat Forces Assessment Model (CFAM), a large-scale linear programming model for weapons allocation analyses. The OptQuest Callable Library (OCL) is described in detail in Chapter 9. Three notional planning scenarios were used and a robust solution sought to the multi-scenario problem.   He compared OptQuest results with previous results obtained using a genetic algorithm (for the same scenarios). His results indicated that better overall solutions, a greater diversity of solutions, and quicker convergence results were obtained using the OptQuest scatter search approach.

The methodology proposed by Hill and McIntyre (2000) and implemented by Bulut (2001) using OptQuest is directly applicable to any analytical situation involving competing "scenarios" within which one needs a single solution that is "robust" relative to any differences among the scenarios.

Chapter 9

COMMERCIAL SCATTER SEARCH IMPLEMENTATION

Everything should be made as simple as possible, but not simpler.

Albert Einstein

In this chapter we discuss the development of commercial optimization software based on the scatter search methodology. The OptQuest Callable Library (OCL), which we began developing in the fall of 1998 in collaboration with Fred Glover and James P. Kelly, is the optimization engine of the OptQuest system[1]. The main goal of OptQuest is to optimize complex systems, which we consider to be those that cannot be easily formulated as mathematical models and solved with classical optimization tools. Many real world optimization problems in business, engineering and science are too complex to be given tractable mathematical formulations. Multiple nonlinearities, combinatorial relationships and uncertainties often render challenging practical problems inaccessible to modeling except by resorting to more comprehensive tools (like computer simulation). Classical optimization methods encounter grave difficulties when dealing with the optimization problems that arise in the context of complex systems. In some instances, recourse has been made to itemizing a series of scenarios in the hope that at least one will give an acceptable solution. Due to the limitations of this approach, a long-standing research goal has been to create a way to guide a series of complex evaluations to produce high quality solutions, in the absence of tractable mathematical structures. (In the context of optimizing simulations, a "complex evaluation" refers to the execution of a simulation model given a set of values for key input parameters.)

---

[1] OptQuest is a registered trademark of OptTek Systems, Inc. (www.opttek.com). The descriptions in this chapter are based on OCL 4.0.

Theoretically, the issue of identifying best values for a set of decision variables falls within the realm of optimization. Until quite recently, however, the methods available for finding optimal decisions have been unable to cope with the complexities and uncertainties posed by many real world problems of the form treated by simulation. The area of stochastic optimization has attempted to deal with some of these practical problems, but the modeling framework limits the range of problems that can be tackled with such technology.

The complexities and uncertainties in complex systems are the primary reason that simulation is often chosen as a basis for handling the decision problems associated with those systems. Consequently, decision makers must deal with the dilemma that many important types of real world optimization problems can only be treated by the use of simulation models, but once these problems are submitted to simulation there are no optimization methods that can adequately cope with them.

Recent developments are changing this picture. Advances in metaheuristics such as scatter search have led to the creation of optimization engines that can successfully guide a series of complex evaluations with the goal of finding optimal values for the decision variables. One of those engines is the search procedure embedded in OCL.

OCL is designed to search for optimal solutions to the following class of optimization problems:

$$\text{Max or Min} \quad f(x,y)$$

Subject to	$Ax \le b$	(Constraints)
	$g_l \le g(x,y) \le g_u$	(Requirements)
	$l \le x \le u$	(Bounds)
	$y = \text{alldifferent}$	

where x can be continuous or discrete with an arbitrary step size and y represents a permutation.

The objective $f(x,y)$ may be any mapping from a set of values x and y to a real value. The set of constraints must be linear and the coefficient matrix (A) and the right-hand-side values (b) must be known. The requirements are simple upper and/or lower bounds imposed on a function that can be linear or non-linear. The values of the g_l and g_u bounds must be known constants. All the variables must be bounded and some may be restricted to be discrete with an arbitrary step size. The set of y variables are used to represent permutations. A given optimization model may consist of any combination of continuous, discrete or "alldifferent" (also referred to as "permutation") variables.

In a general-purpose optimizer such as OCL, it is desirable to separate the solution procedure from the complex system to be optimized. The disadvantage of this "black box" approach is that the optimization procedure is generic and has no knowledge of the process employed to perform evaluations inside of the box and therefore does not use any problem-specific information (Figure 9-1). The main advantage of this design, on the other hand, is that the same optimizer can be applied to complex systems in many different settings.

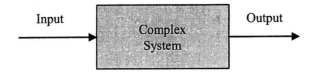

Figure 9-1. Complex system as a black box

OCL is a generic optimizer that overcomes the deficiency of black box systems of the type illustrated in Figure 9-1, while successfully embodying the principle of separating the method from the model. In such a context, the optimization problem is defined outside the complex system. Therefore, the evaluator can change and evolve to incorporate additional elements of the complex system, while the optimization routines remain the same. Hence, there is a complete separation between the model used to represent the system and the procedure that solves optimization problems based on the model.

The optimization procedure uses the outputs from the system evaluator, which measures the merit of the inputs that were fed into the model. On the basis of both current and past evaluations, the optimization procedure decides upon a new set of input values (see Figure 9-2).

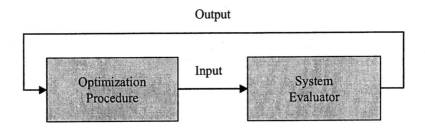

Figure 9-2. Coordination between optimization and system evaluation

The optimization procedure is designed to carry out a special "strategic search," where the successively generated inputs produce varying evaluations, not all of them improving, but which over time provide a highly efficient trajectory to the best solutions. The process continues until an appropriate termination criterion is satisfied (usually based on the user's preference for the amount of time to be devoted to the search).

OCL enables the development of applications to solve optimization problems using the "black-box" approach for evaluating an objective function and a set of requirements. Figure 9-3 shows a conceptualization of how OCL can be used to search for optimal solutions to complex optimization problems.

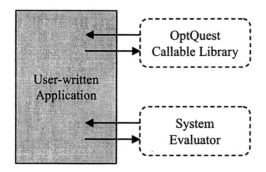

Figure 9-3. OCL linked to a user-written application

Figure 9-3 assumes that the user has a system evaluator that, given a set of input values, it returns a set of output values that can used to guide a search. For example, the evaluator may have the form of a computer simulation that, given the values of a set of decision variables, it returns the value of one or more performance measures (that define the objective function and possibly a set of requirements). The user-written application uses OCL functions to define an optimization problem and launch a search for the optimal values of the decision variables.

A simple example in risk analysis is to use a Monte Carlo simulator to estimate the average rate of return and the corresponding variance of a proposed investment portfolio. The average rate of return may be used to define the objective function and the estimated variance may be used to formulate a requirement to limit the variability of the returns. The decision variables represent the allocation of funds in each investment alternative and a linear constraint can be added to limit the total amount invested. In this case, the user must provide the simulator to be coupled with OCL, while the optimization model is entirely formulated with calls to the appropriate library functions.

1.   GENERAL OCL DESIGN

OCL seeks to find an optimal solution to a problem defined on a vector x of bounded variables and/or a permutation y. That is, the user can define several types of optimization problems depending on the combination of the variable types:

- Pure continuous
- Pure discrete (including pure binary problems)
- Pure permutation ("all different" variables only)
- Mixed problems (continuous-discrete, continuous-permutation, discrete-permutation or continuous-discrete-permutation)

Also, the optimization problem may be unconstrained, include linear constraints and/or requirements. Hence, OCL can be used to formulate up to 28 different types of problems. OCL detects small pure discrete or pure permutation problems to trigger a complete enumeration routine that guarantees optimality of the best solution found.

In this chapter, we will describe the mechanisms that OCL employs to search for optimal solutions to problems defined with continuous and discrete variables. Similar mechanisms are used to tackle pure or mixed permutation problems, such as those described in Chapter 7. The scatter search method implemented in OCL uses the Diversification Generation Method described in Section 1 of Chapter 2. In addition to the solutions generated with the Diversification Generation Method, the initial set P includes the following solutions:

- All variables are set to the lower bound
- All variables are set to the upper bound
- All variables are set to the midpoint $x = l + (u - l)/2$
- Other solutions suggested by the user

A subset of diverse solutions is chosen as members of the reference set. A set of solutions is considered diverse if its elements are "significantly" different from one another. OCL uses a Euclidean distance measure to determine how "close" a new trial solution is from the solutions already in the reference set, in order to decide whether the new trial solution is included or discarded. (See Section 3 of Chapter 2 for details about the distance measure used for continuous and discrete variables.)

When the optimization model includes discrete variables, a rounding procedure is used to map fractional values to discrete values. When the model includes linear constraints new trial solutions are subjected to a feasibility test before they are sent to the evaluator (i.e., the feasibility check is performed before the objective function value $f(x)$ and the requirements $g(x)$ are evaluated). Note that the evaluation of the objective function may entail the execution of a simulation, and therefore it is important to be sure to evaluate only those solutions that are feasible with respect to the set of constraints. For ease of notation, we represent the set of constraints as $Ax \leq b$, although equality constraints are also allowed. The feasibility test consists of checking (one by one) whether the linear constraints are satisfied. If the solution is infeasible with respect to one or more constraints, OCL formulates and solves a linear programming (LP) problem. The LP (or mixed-integer program, when x contains discrete variables) has the goal of finding a feasible solution x^* that minimizes a deviation between x and x^*. Mathematically, the problem can be formulated as:

$$\text{Minimize} \quad d^- + d^+$$

$$\begin{aligned}
\text{subject to} \quad & Ax^* \leq b \\
& x - x^* - d^- + d^+ = 0 \\
& l \leq x^* \leq u \\
& d^-, d^+ \geq 0
\end{aligned}$$

where d^- and d^+ are, respectively, negative and positive deviations of x^* from the infeasible solution x. The implementation of this mechanism within OCL includes a scaling procedure to account for the relative magnitude of the variables and adds a term to the objective function to penalize maximum deviation. Also, OCL treats pure binary problems differently, penalizing deviations without adding deviation variables or constraints. When the optimization problem does not include constraints, infeasible solutions are made feasible by simply adjusting variable values to their closest bound and rounding when appropriate. That is, if $x > u$ then $x^* = u$ and if $x < l$ then $x^* = l$.

Once the reference set has been created, a Combination Method is applied to initiate the search for optimal solutions. The method consists of finding linear combinations of reference solutions. The combinations are based on the following three types, which assume that the reference solutions are x' and x'':

$$x = x' - d$$
$$x = x' + d$$
$$x = x'' - d$$

where $d = r \dfrac{x'' - x'}{2}$ and r is a random number in the range $(0, 1)$. Because a different value of r is used for each element in x, the Combination Method can be viewed as a sampling procedure in a rectangle instead of a line in a two dimensional space, as shown in Section 3 of Chapter 5. The number of solutions created from the linear combination of two reference solutions depends on the quality of the solutions being combined. Specifically, when the best two reference solutions are combined, they generate up to 5 new solutions, while when the worst two solutions are combined they generate only one. This follows the strategy described in Section 3.1 of Chapter 5.

In the process of searching for a global optimum, the Combination Method may not be able to generate solutions of high enough quality to become members of the reference set. If the reference set does not change and all the combinations of solutions have been explored, a diversification step is triggered. This step consists of rebuilding the reference set to create a balance between solution quality and diversity. To preserve quality, a small set of the best (*elite*) solutions in the current reference set is used to seed the new reference set. The remaining solutions are eliminated from the reference set. Then, the Diversification Generation Method is used to repopulate the reference set with solutions that are diverse with respect to the elite set. This reference set is used as the starting point for a new round of combinations. The rebuilding mechanism is the same as the one described in Section 1.2.1 of Chapter 5.

2. CONSTRAINTS AND REQUIREMENTS

So far, we have assumed that the complex system to be optimized can be treated by OCL as a "black box" that takes x as input to produce $f(x)$ as output. We have also assumed that for x to be feasible, the solution must be within a given set of bounds and, when applicable, also satisfy a set of linear constraints. We assume that both the bounds and the coefficient matrix are known. However, there are situations where the feasibility of x is not known prior to performing the process that evaluates $f(x)$, i.e., prior to executing the "black box" system evaluator. In other words, the feasibility test for x cannot be all performed in the input side of the black box but instead the black box has to evaluate $g(x)$ and the complete feasibility of x must be tested in the output side. This situation is depicted in Figure 9-4.

Figure 9-4. Solution evaluation

Figure 9-4 shows that when constraints are included in the optimization model, the evaluation process starts with the mapping $x \rightarrow x^*$. If the only constraints in the model are integrality restrictions, the mapping is achieved with a rounding mechanism that transforms fractional values into integer values for the discrete variables. If the constraints are linear, then the mapping consists of formulating and solving the abovementioned linear programming problem. Finally, if the constraints are linear and the model also includes discrete variables, then the linear programming formulation becomes a mixed-integer programming problem that is solved accordingly. Clearly, if the optimization model has no constraints or discrete variables then $x^* = x$.

The complex system evaluator uses the mapped solution x^* to obtain a set of performance measures (i.e., the output of the evaluation). One of these measures is used as the objective function value $f(x)$ and provides the means for the search to distinguish high-quality from inferior solutions. Other measures $g(x)$ associated with the performance of the system can be used to define a set of requirements. A requirement is expressed as a bound on the value of a performance measure $g(x)$. Thus, a requirement may be defined as an upper or a lower bound on an output of the complex system evaluator. Instead of discarding requirement-infeasible solutions, OCL handles them with a composite function $p(x)$ that penalizes the requirement violations. The penalty is proportional to the degree of the violation and is not static throughout the search. OCL assumes that the user is interested in finding a requirement-feasible solution if one exists. Therefore, requirement-infeasible solutions are penalized more heavily when no requirement-feasible solution has been found during the search than when one is already available. Also, requirement-feasible solutions are always considered superior to requirement-infeasible solutions.

To illustrate the evaluation process in the context of a simulated system, let us revisit the investment problem briefly introduced earlier in this chapter. In this problem, x represents the allocation of funds to a set of investment instruments. The objective is to maximize the expected return. Assume that a Monte Carlo simulation is performed to estimate the expected return $f(x)$ for a given fund allocation. Hence, in this case, the complex system evaluator consists of a Monte Carlo simulator.

Restrictions on the fund allocations, which establish relationships among the x variables, are handled within the linear programming formulation that maps infeasible solutions into feasible ones. Thus, a restriction of the type "the combined investment in instruments 2 and 3 should not exceed the total investment in instrument 7," results in the linear constraint $x_2 + x_3 \leq x_7$. On the other hand, a restriction that limits the variability of the returns (as measured by the standard deviation) to be no more than a critical value c cannot be enforced in the input side of the Monte Carlo simulator. Clearly, the simulation must be executed first in order to estimate the variability of the returns. Suppose that the standard deviation of the returns is represented by $g(x)$ then the requirement in this illustrative situation is expressed as $g(x) \leq c$.

Note that the constraint-mapping mechanism within OCL does not handle nonlinear constraints. However, nonlinear constraints can be modeled as requirements and incorporated within the penalty function $p(x)$. For example, suppose that an optimization model must include the following nonlinear constraint:

$$x_1 x_2 - x_1^2 \leq 120$$

Then, the complex system evaluator calculates, for a given solution x, the left-hand side of the nonlinear constraint and communicates the result as one of the outputs. OCL uses this output and compares it to the right-hand side value of 120 to determine the feasibility of the current solution. If the solution is not feasible a penalty term is added to the value of the objective function $f(x)$.

3. OCL FUNCTIONALITY

The OptQuest Callable Library consists of a set of 47 functions that are classified into 12 categories:

- Start and Ending Optimization
- Null Values
- Variable Definition
- Constraint Definition
- Requirement Definition
- Solutions
- Parameters
- Event Handling
- Information Retrieval

- Mixed Integer Programming
- Internal Functions
- Neural Network

The functions are classified according to their purpose and a full description of each of them is provided in the OptQuest User's Manual located in the Documentation-C folder of the Chapter9 directory. The documentation is web-based and is accessed by double-clicking on OptQuestManual.html.

The main OCL functions are used to setup an optimization problem and perform the scatter search. Additional functions in the library are used to change parameter settings or perform advanced operations such as monitoring and changing the composition of the reference set. The library also allows the user to define and train a neural network as well as to define and solve a linear (or mixed integer) programming problem. Regardless of the complexity of the application that uses OCL as its optimization engine, the following structure is generally followed:

- Allocate memory for the optimization model by indicating the number of variables, constraints and requirements in the problem, as well as defining the direction of the optimization as minimize or maximize (`OCLSetup`).
- Define continuous and/or discrete decision variables (`OCLDefineVar`).
- Initialize the reference set (`OCLInitPop`) or generate all solutions in the case of small pure discrete problems (`OCLGenerateAllSolutions`).
- Iterate by retrieving a solution from OCL's database (`OCLGetSolution`), evaluating the solution (user-provided system evaluator) and placing the evaluated solution back into OCL's database (`OCLPutSolution`).

Suppose that we would like to use OCL to search for the optimal solution to the following unconstrained nonlinear optimization problem:

$$\text{Minimize} \quad 100\left(x_2 - x_1^2\right)^2 + \left(1 - x_1\right)^2 + 90\left(x_4 - x_3^2\right)^2 + \left(1 - x_3\right)^2 +$$
$$10.1\left(\left(x_2 - 1\right)^2 + \left(x_4 - 1\right)^2\right) + 19.8\left(x_2 - 1\right)\left(x_4 - 1\right)$$

$$\text{Subject to} \quad -10 \le x_i \le 10 \quad \text{for } i = 1, \ldots, 4$$

This is the same problem that we used in our tutorial Chapter 2. It is also the same problem in the folder Examples-C\Example1 of the Chapter9 directory. According to the general structure of OCL, we need to start by allocating memory and indicating the direction of the optimization. To do this, we use the OCLSetup function, which has the following prototype:

```
long OCLSetup(long nvar, long nperm, long ncons, long
        req, char *direc, long lic);
```

nvar	An integer indicating the number of continuous and discrete decision variables in the problem
nvar	An integer indicating the number of permutation ("all different") decision variables in the problem
ncons	An integer indicating the number of constraints in the problem
req	An integer indicating the number of requirements in the problem
direc	An array of characters with the characters "MAX" to indicate maximization or "MIN" to indicate minimization
lic	A valid license number

Therefore, the OCLSetup function call for our example would look like this:

```
nprob = OCLSetup(4, 0, 0 , 0, "MIN", 999999999);
```

where nprob is a positive integer that uniquely identifies the problem within OCL's memory. If OCLSetup returns a negative value, then the setup operation has failed. The license code shown above is the one distributed with the demo version of OCL. The demo version is the same as the full version with the exception of the limitation on the size of the model (i.e., number of variables, constraints and requirements) and the limit on the number of iterations. After setting up the problem, we need to define the decision variables using the OCLDefineVar function that has the following prototype:

```
long OCLDefineVar(long nprob, long var, double low,
        double sug, double high, char *type, double step);
```

nprob	A unique number that identifies an optimization problem within OCL's memory. This is the identifier returned by OCLSetup.
var	An integer indicating the variable number that corresponds to the current definition.
low	A double indicating the minimum value for the corresponding variable.
sug	A double indicating the suggested value for the corresponding variable. The suggested value is typically included in the initial reference set, unless the value results in an infeasible solution.

The OCLNULL value can be used when no suggested value is available. The OCLNULL value is used to ignore an argument of an OCL function and is obtained with a call to OCLGetNull.

high   A double indicating the maximum value for the corresponding variable.

type   An array of characters with the word "CON" to define a continuous variable or "DIS" to define a discrete variable.

step   A double indicating the step size for a discrete variable. Step sizes may be integer or fractional and must be strictly greater than zero. Step sizes for continuous variables are ignored.

The function call to define the variables in our example can be programmed as follows:

```
for (i = 1; i <= 4; ++i)
 OCLDefineVar(nprob,i,-10,OCLGetNull(nprob),10,"CON",1);
```

Note that although we use a "1" as the last argument of the function, this value is ignored because all the variables are defined as continuous.

We are now ready to build the starting reference set. This step is performed with a call to the OCLInitPop function. The "Pop" in the function name refers to "population", which is the terminology preferred by the GA community, although the term "reference set" is more common in scatter search implementations. The prototype of this function consists of a single argument containing the unique problem identifier returned by OCLSetup. The function call looks as follows, which assumes that nprob is the identifier returned by OCLSetup:

```
OCLInitPop(nprob);
```

It is important to point out that all the functions in OCL return an integer value. If the return value is positive, the function call was successful. Otherwise, if the return value is negative, the function call failed and the return value is the error code. A list of error codes can be found in the documentation.

After a successful initialization of the reference set, the search can begin. The search is performed with a series of calls to three functions: OCLGetSolution, a user-provided system evaluator and OCLPutSolution. The first function retrieves a solution from OCL's database, the second evaluates the solution and the third places the evaluated solution back into OCL's database. The prototype for OCLGetSolution is:

```
long OCLGetSolution(long nprob, double *sol);
```

sol   A pointer to an array of doubles where OCL places the solution. The array should have enough space to hold the variable values in positions 1 to `nvar`, as defined in `OCLSetup`.

As before, `nprob` is the unique problem identifier returned by `OCLSetup`. If the call to `OCLGetSolution` is successful, the function returns a unique solution identifier `nsol`. The prototype for `OCLPutSolution` is:

```
long OCLPutSolution(long nprob, long nsol, double
      *objval, double *sol);
```

nsol   A solution identifier returned by `OCLGetSolution`.

objval   A pointer to a double where the objective function value is stored or an array of doubles with the values of the objective function and the requirements. The array should have a size of at least `req+1` positions, as defined in `OCLSetup`. The objective function value should be `objval[0]` and the ith requirement should be `objval[i]`. Note that if no requirements are defined, then `objval` can be dimensioned as a simple double variable. `OCLNULL` may be used to instruct OCL to discard the solution.

sol   An array of doubles with the values of the decision variables. If the solution values are the same as the ones retrieved with a call to `OCLGetSolution`, a pointer to the `OCLNULL` variable may be used.

The evaluation function is outside the scope of OCL and is the responsibility of the user. For our example, we can code the evaluator simply as:

```
double evaluate(double *x)
{
        return(100*pow(x[2]-pow(x[1],2),2)+pow(1-x[1],2)
               +90*pow(x[4]-pow(x[3],2),2)+pow(1-x[3],2)
               +10.1*((pow(x[2]-1,2)+pow(x[4]-1,2))
               +19.8*(x[2]-1)*(x[4]-1));
}
```

Assuming that we want to search for the optimal solution to this problem allowing OCL to perform a maximum of 10000 function evaluations, the code to perform such a search is the one shown in Figure 9-5.

```
for (i = 1; i <= 10000; i++)
{
  nsol = OCLGetSolution(nprob, x);
  if (nsol < 0) {
    printf("OCLGetSolution error code %d\n", nsol);
    exit(1);
  }
  objval = evaluate(x);
  status = OCLPutSolution(nprob, nsol, &objval,
          (double *)OCLGetNull(nprob));
  if (status < 0) {
    printf("OCLPutSolution error code %d\n", status);
    exit(1);
  }
}
```

Figure 9-5. Search loop in example1.c

The code in Figure 9-5, evaluates and returns the objective function value of 10000 solutions. It also checks for possible error codes from calls to OCLGetSolution and OCLPutSolution. Note that the partial code in Figure 9-5 does not keep track of the best solution found. This can be done by adding an "if" statement that compares the objective function value of the current solution with the best objective function value found during the search. Alternatively, the OCLGetBest function can be called at any time during the search to retrieve the values associated with the best solution, which OCL automatically monitors. This function is called with the following arguments:

```
OCLGetBest(nprob, x, &objval);
```

where nprob is the unique problem identifier, x is the array where the variable values are stored and objval is the variable where the objective function value is returned. The entire C code for this example is example1.c in the folder Chapter9\Examples-C\Example1 of the accompanying disc.

3.1   Defining Constraints and Requirements

The illustration in the previous section did not include the OCL functions for constraints and requirements. In this section, we briefly describe how constraints and requirements can be defined with OCL. Assume that we would like to add the following linear constraint to the optimization model for our 4-variable example problem:

$$x_1 + 8x_2 - 3x_4 \leq 5$$

Then after a call to OCLSetup and before a call to OCLInitPop, we must add calls to the constraint-related functions: OCLConsCoeff (to change the coefficient of a constraint), OCLConsRhs (to change the right-hand-side of a constraint) and OCLConsType (to change the constraint type). The calls to these functions can be made in any order, as long as they are made before the reference set is initialized and certainly before the search begins with calls to OCLGetSolution and OCLPutSolution. The prototypes for the constraint-related functions are:

```
long OCLConsCoeff(long nprob, long cons, long var,
        double coeffval);
long OCLConsRhs(long nprob, long cons, double rhsval);
long OCLConsType(long nprob, long cons, long type);
```

In these prototypes, nprob is the unique problem identifier returned by OCLSetup, cons is the constraint number, var is the variable number, coeffval is the coefficient value, rhsval is the right-hand-side value and type is the constraint type (OCLLE = less-than-or-equal, OCLGE = greater-than-or-equal, and OCLEQ = equal). The following function calls can be used to define the constraint in our example:

```
OCLConsCoeff(nprob, 1, 1,  1);
OCLConsCoeff(nprob, 1, 2,  8);
OCLConsCoeff(nprob, 1, 4, -3);
OCLConsRhs(nprob, 1, 5);
OCLConsType(nprob, 1, 1, OCLLE);
```

These function calls assume that the constraint is the first one in the model. Also, the variable numbers match the ones used when OCLDefineVar was called. A complete illustration of defining constraints in OCL can be found in the Chapter9\Examples-C\Example2 folder of the accompanying disc.

The definition of requirements is slightly different from the definition of constraints. As mentioned before, a requirement is basically a bound on an output value of the system evaluator. Suppose that we would like to define a requirement to impose the following nonlinear restriction to our illustrative example:

$$10.8x_1^2 - 2.4x_3x_4 \geq 15.9$$

We first use a call to OCLDefineReq to define the requirement. This function has the following prototype:

```
long OCLDefineReq(long nprob, long req, double low,
        double high);
```

Where nprob is the unique problem identifier returned by OCLSetup, req is the requirement number, low is the lower bound for the requirement, and high is the upper bound for the requirement. The variable OCLNULL can be used to leave either low or high undefined. The function call for our example is:

```
OCLDefineReq(nprob, 1, 15.9, OCLGetNull(nprob));
```

In addition to this definition, we need to modify the system evaluator. The evaluator must return the value of the objective function in objval[0] and the value of the requirement in objval[1]. The new evaluator looks like this:

```
void evaluate(double *x, double *objval)
{
   objval[0] = 100*pow(x[2]-pow(x[1],2),2)+pow(1-x[1],2)
               +90*pow(x[4]-pow(x[3],2),2)+pow(1-x[3],2)
               +10.1*((pow(x[2]-1,2)+pow(x[4]-1,2))
               +19.8*(x[2]-1)*(x[4]-1));
   objval[1] = 10.8*pow(x[1],2)-2.4*x[3]*x[4];
}
```

Other small changes are necessary to make OCL work with the requirement that we have defined. For example, the declaration of objval must be changed from a single double to an array of doubles and the evaluate function must be changed from double to void. The changes are reflected in the following partial code, where TotalIter is the total number of objective function evaluations to be executed during the search:

```
void evaluate(double *, double *);
.
.
.
   double objval[3];
```

```
for (i = 1; i <= TotalIter; i++) {
  nsol = OCLGetSolution(nprob, x);
  evaluate(x, objval);
  OCLPutSolution(nprob, nsol, objval, x);
}
```

When requirements are included in an optimization model, OCLGetBest indicates the feasibility of the best solution. The return values for OCLGetBest can be either 0 if the best solution is requirement-feasible, 1 if the best solution is requirement-infeasible or a negative value representing an error code. Recall that while all the solutions generated by OCL are constraint-feasible (if the feasibility region is not empty), some may be requirement-infeasible and this is why the OCLGetBest function is designed to provide information regarding the requirement-feasibility of the best solution found during the search.

The discussion about the functionality of OCL in which we have engaged is not meant to be exhaustive. OCL has additional functions that can be used to modify the way the search is performed. For example, OCL includes functions to change the default parameter settings. A function such as OCLSetSearchStrategies can be used to change the parameter settings of strategies that define the search trajectory. This function is meant for the advanced user, who has a thorough understanding of both the problem context and the way the search method works.

3.2 Boundary Search Strategy

The OCLSetSearchStrategies function is used to change several settings of several user-controlled search strategies. One of these strategies is known as the boundary search strategy, which is a mechanism to generate non-convex combinations of two solutions. In particular, the set of points generated on the line defined by x' and x'', but beyond the segment $[x', x'']$ "outside" of x'' is given by:

$$x = x'' + (x'' - x')t \qquad \text{for } t > 0$$

While the "standard" combination method in OCL defines $t = r/2$, where r is a random number between 0 and 1, the boundary strategy considers three strategies to generate non-convex combinations in the (x', x'') line:

- *Strategy* 1: Computes the maximum value of t that yields a feasible solution x when considering both the bounds and the constraints in the model. An additional solution is also generated that is halfway between

x'' and the solution at the boundary. This strategy is illustrated in Figure 9-6 for an unconstrained (but bounded) problem. The constrained case is depicted in Figure 9-7. In these figures, points 1 and 2 represent the reference solutions x' and x'', while point 3 is the one reached with the maximum value of t and point 4 is the additional halfway solution. Also in these figures, the shaded area represents the feasible region.

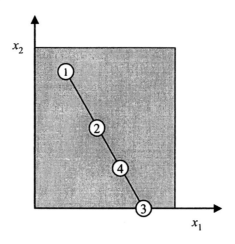

Figure 9-6. Boundary strategy 1 (unconstrained case)

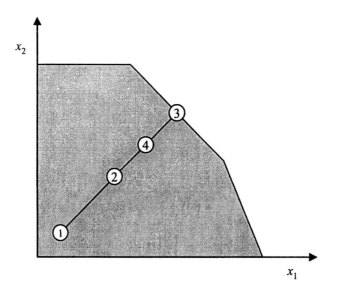

Figure 9-7. Boundary strategy 1 (constrained case)

- *Strategy* 2: Considers the fact that variables may hit bounds before leaving the feasible region relative to other constraints. The first departure variable from the feasible region may happen because some variable hits a bound, which is followed by the others, before any of the linear constraints is violated. In such a case, the departing variable is fixed at its bound when it hits it, and the exploration continues with this variable held constant. OCL does this with each variable that encounters a bound before other constraints are violated. The process finishes when the boundary defined by the other constraints is reached. Figure 9-8 and 9-9 depict this strategy for the unconstrained and constrained cases, respectively. In both cases, the line that joins the reference points 1 and 2 is extended until hitting a variable bound, then the search moves on the boundary until hitting the bound corresponding to the other variable. The extreme point 3 and point 4 that is halfway between points 2 and 3 are generated.

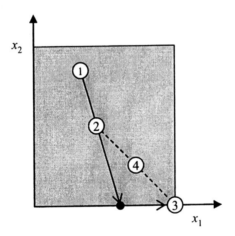

Figure 9-8. Boundary strategy 2 (unconstrained case)

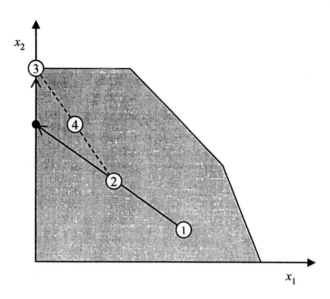

Figure 9-9. Boundary strategy 2 (constrained case)

– *Strategy* 3: Considers that the exploration hits a boundary that may be
   defined by either bounds or any of the linear constraints. When this
   happens, one or more constraints may be binding and the corresponding
   t-value cannot be increased without causing the violation of at least one
   constraint. At this point, OCL chooses a variable to make a substitution
   that geometrically corresponds to a projection that makes the search
   continue on a line that has the same direction, relative to the constraint
   that was reached, as it did upon approaching the hyperplane defined by
   the constraint. The process continues until the last unfixed variable hits a
   constraint. At this point, the value of all the previously fixed variables is
   computed. Figure 9-10 depicts this strategy.

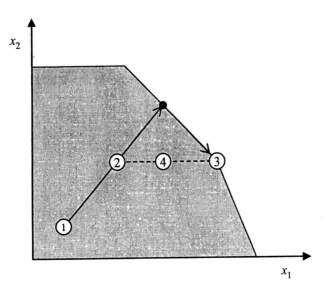

Figure 9-10. Boundary strategy 3

Each of these three boundary strategies generates a "boundary solution" x^b outside x''. OCL also generates the solution in the midpoint between x^b and x''. Interchanging the role of x' and x'' gives the extension outside the "other end" of the line segment. The mechanism results in a total of 4 non-convex solutions out of a single combination of a pair of reference solutions. The user establishes, by setting the appropriate value in the OCLSetSearchStrategies function, the percentage of combinations in which the boundary strategy is applied. The default value calls for applying the boundary strategy 50% of the time. Although this default value is fairly robust across a variety of problem classes, for problems where the best values are suspected to lie near the boundary of the feasible region, improved solutions may be found when the frequency value is increased. Similarly, if the best solutions to an optimization problem are not near a boundary, improved outcomes may result from decreasing the frequency for applying the boundary strategy.

4.   COMPUTATIONAL EXPERIMENTS

In this section, we apply the OptQuest Callable Library to a set of hard nonlinear and unconstrained optimization problems. We have built an application based on OCL that allows us to test the performance of the library as compared to a competing generic optimizer based on genetic algorithms. To perform the experiments, we have used a fully-licensed version of OptQuest, which imposes no limits on the size of the optimization problem. The reader will be able to reproduce the experiments for problem instances with up to 5 variables, which is the limit imposed on the demo version of OptQuest included in this book.

Table 9-1 shows the set of 40 test problems that we have collected from several sources, including the following web pages:

http://www.maths.adelaide.edu.au/Applied/llazausk/alife/realfopt.htm
http://solon.cma.univie.ac.at/~neum/glopt/my_problems.html
http://www-math.cudenver.edu/~rvan/phd/node32.html

Table 9-1. Test problems

Number of variables	Problem number	Name and parameter values	x^*	$f(x^*)$
2	1	Branin	(9.42478, 2.475)†	0.39789
	2	B2	(0, 0)	0
	3	Easom	(π,π)	-1
	4	Goldstein and Price	(0, -1)	3
	5	Shubert	(0.0217, -0.9527)†	-186.731
	6	Beale	(3, 0.5)	0
	7	Booth	(1, 3)	0
	8	Matyas	(0, 0)	0
	9	SixHumpCamelback	(0.089840, -0.712659) †	-1.03163
	10	Schwefel(2)	(1, 1)	0
	11	Rosenbrock (2)	(1, 1)	0
	12	Zakharov(2)	(0, 0)	0
3	13	De Joung	(0, 0, 0)	0
	14	Hartmann(3,4)	(0.114614, 0.555649, 0.852547)	0
4	15	Colville	(1, 1, 1, 1)	0
	16	Shekel(5)	(4, 4, 4, 4)	-10.1532
	17	Shekel(7)	(4, 4, 4, 4)	-10.4029
	18	Shekel(10)	(4, 4, 4, 4)	-10.5364
	19	Perm(4,0.5)	(1, 2, 3, 4)	0
	20	Perm0(4,10)	(1, 1/2, 1/3, 1/4)	0
	21	Powersum (8,18,44,114)	(1, 2, 2, 3)	0
6	22	Hartmann(6,4)	(0.20169, 0.150011, 0.47687, 0.275332, 0.311652, 0.6573)	0
	23	Schwefel(6)	(1, ..., 1)	0
	24	Trid(6)	$x_i = i*(7-i)$	-50
10	25	Trid(10)	$x_i = i*(11-i)$	-210
	26	Rastrigin(10)	(0, ..., 0)	0
	27	Griewank(10)	(0, ..., 0)	0
	28	Sum Squares(10)	(0, ..., 0)	0
	29	Rosenbrock(10)	(1, ..., 1)	0
	30	Zakharov(10)	(0, ..., 0)	0

Number of variables	Problem number	Name and parameter values	x^{*}	$f(x^{*})$
20	31	Rastrigin(20)	$(0, ..., 0)$	0
	32	Griewank(20)	$(0, ..., 0)$	0
	33	Sum Squares(20)	$(0, ..., 0)$	0
	34	Rosenbrock(20)	$(1, ..., 1)$	0
	35	Zakharov(20)	$(0, ..., 0)$	0
> 20	36	Powell(24)	$(3, -1, 0, 1, 3, ..., 3, -1, 0, 1)$	0
	37	Dixon and Price(25)	$x_i = 2^{-((z-1)/z)}, z = 2^{i-1}$	0
	38	Levy(30)	$(1, ..., 1)$	0
	39	Sphere(30)	$(0, ..., 0)$	0
	40	Ackley(30)	$(0, ..., 0)$	0

† This is one of multiple optimal solutions.

The numbers between parentheses associated with some of the function names in Table 9-1 are the parameter values for the corresponding objective function. A typical parameter refers to the number of variables in the function, since many of these functions expand to an arbitrary number of variables. Although the objective functions are built in a way that the optimal solutions are known, the optimization problems cannot be trivially solved by search procedures that do not exploit the special structure that characterizes each function. The optimal solutions and the corresponding objective function values are shown on the last two columns of Table 9-1. A detailed description of the objective functions is provided in the Appendix at the end of this chapter.

Function 9-1 shows the main function of our OCL-based code for the solution of the nonlinear optimization problems in Table 9-1. This function has three arguments: the problem number, the total number of function evaluations and the setting for the "aggressive" search strategy. The problem number is used to select the evaluation of the correct objective function from the catalog of functions in the `evaluate` function, which is in the evaluate.c file along with the `n_var` function. The `n_var` function is called first to retrieve the number of variables in the problem, the function name, the bound values and the optimal objective function value.

Function 9-1. main — File: oclmain.c

```
int main(int argc, char **argv)
{
   double  objval, x[100];
   long    nprob, nsol;
   int     i, prob, nvar, TotalIter;
   double  lower, upper, optval, pval;
   char    fname[50];

   if (argc != 4)
```

```
{
   printf("usage: ocl <iter> <prob> <aggressive>\n");
   exit(1);
}
TotalIter = atoi(argv[1]);
prob = atoi(argv[2]);
pval = atof(argv[3]);

nvar = n_var(prob,fname,&lower,&upper,&optval);
nprob = OCLSetup(nvar,0,0,0,"MIN",999999999);
for(i=1;i<=nvar;i++)
OCLDefineVar(nprob,i,lower,OCLGetNull(nprob),upper,"CON",1);
OCLSetSearchStrategies(nprob,"aggressive",pval);
OCLInitPop(nprob);
for (i = 1; i <= TotalIter; i++)
{
   nsol=OCLGetSolution(nprob,x);
   objval = evaluate(prob,nvar,x);
   OCLPutSolution(nprob,nsol,&objval,
   (double *)OCLGetNull(nprob));
   if(!(i%100))
   {
      OCLGetBest(nprob,x,&objval);
      printf("Iter: %6d  Best ObjVal: %15.6lf\n",i,objval);
   }
}
OCLGoodBye(nprob);
return 0;
}
```

Although we are not using the function name or the optimal objective function value in Function 9-1, this information can be used to generate an output table with the relative deviation of the solution found by OCL from the optimal solution. The OCLSetSearchStrategies function is called to set the "aggressive" strategy on or off, according to the pval argument value. When pval = 1 the aggressive strategy is used. After a call to OCLInitPop, the main for-loop starts. The for-loop includes an if-statement to print, every 100 iterations, the objective function value of the best solution found.

After running OCL for 100 thousand objective function evaluations (i.e., TotalIter = 100000) with the "aggressive" strategy off (i.e., at its default value), the optimizer is capable of finding 33 out of 40 optimal solutions. The average deviation from optimal is 0.8% when problem 23 is ignored. When the "aggressive" strategy is turned off, OCL is not able to find a near-optimal solutions to Problem 23 (Schwefel(6)). However, when the aggressive strategy is used, OCL finds a solution with an objective function value of 0.00301 after 100 thousand iterations. Figure 9-11 shows the percent deviation from the average optimal solution achieved with OCL at 6 points during the search. (This figure ignores the results associated with Problem 23.)

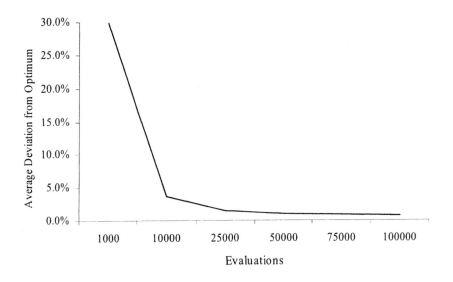

Figure 9-11. OCL search trajectory

The aggressive nature of the scatter search implemented in the OptQuest optimizer is evident in Figure 9-11. Most of the improvement is achieved at the beginning of the search, considering that average optimality gap is 3.7% at 10000-evaluation mark. In terms of CPU time, the largest problems (i.e., those with 30 variables) required 251.8 seconds on a Pentium 4 personal computer at 2.53 GHz. The 2-variable problems required 8.9 seconds on the same machine. The solution time can be reduced if OCL is instructed either to not check for duplicated solutions in its database with a call to `OCLSetCheckDup` or to use a smaller database with a call to `OCLSetDataBaseSize`. The default settings is to check for duplications on a database of 10,000 solutions, because the software is typically used in contexts where the evaluation of the objective function is computationally expensive.

OCL has also been compared to Genocop III[2] on a set of 30 test problems derived from those in Table 9-1 (see Laguna and Martí, 2002). Genocop III is the third generation of a genetic algorithm designed to search for optimal solutions to optimization problems with continuous variables and linear and

---

[2] http://www.coe.uncc.edu/~zbyszek/gchome.html

nonlinear constraints. The description of the first version of Genocop appears in a book by Michalewicz (1996). While this version of Genocop does not handle discrete variables, it does provide a way for explicitly defining nonlinear constraints. The performance of Genocop depends on a set of 12 parameters (without counting the frequency distribution for the application of each operator). For comparison purposes, no attempt was made to find the best parameter setting for the set of problems on hand and instead the default values were used. Similarly, the default values for the OCL parameters were used. The procedures were allowed to perform 10,000 function evaluations, resulting in the following observations:

- OCL found solutions that on the average were better than those found by Genocop, within the scope of the search (i.e., 10000 evaluations).
- OCL found better solutions than Genocop more frequently (20 times for OCL vs. 3 times for Genocop).
- Genocop was on average 4 times faster than OCL.

In addition to comparing the performance of both systems when considering the final solutions and the total computational time to find them, Laguna and Martí (2002) assessed how quickly each method reached the best solutions. As mentioned earlier, reaching good solutions quickly becomes more critical when the complexity of the system evaluator increases. Consider, for example, an application for which a single evaluation of the objective function consists of the execution of a computer simulation that requires 2 CPU minutes. Clearly, in this context, the time to generate each solution becomes negligible. At the same time, it is not feasible to search for the best solution employing 10,000 function evaluations, unless one is willing to wait for 2 weeks to obtain a relatively good answer. A more reasonable approach is to limit the search to 500 function evaluations, whose execution will require somewhat less than 17 hours to execute.

Figure 9-12, from Laguna and Martí (2002), depicts the trajectory followed by OCL and Genocop when tracking the average objective function value in a subset of 28 problems. Note that the graph is drawn on a logarithmic scale to be able to accommodate the large average objective function value of 8.8E+12 yielded by Genocop after 100 evaluations.

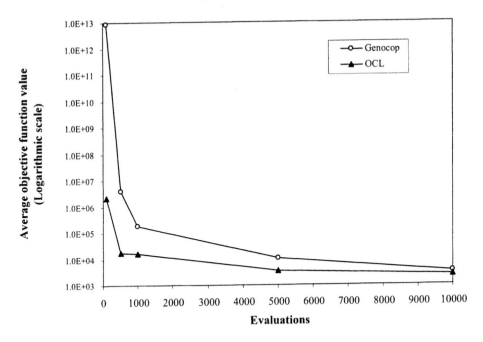

Figure 9-12. Performance graph for OCL and Genocop III

The purpose of constructing the performance graph depicted in Figure 9-12 is to assess the aggressiveness of each procedure as measure by the speed in which the search is capable of finding reasonably good solutions. Based on these results, OCL seems to be more aggressive than Genocop. The aggressiveness in OCL, however, does not compromise the quality of the final solution. In other words, OCL aggressively attempts to improve upon the best solution during the early stages of the search and at the same time it makes use of diversifying mechanisms to be able to sustain an improving trajectory when allowed to extend the search beyond a limited number of objective function evaluations.

The outcomes obtained with OCL for the class of problems tested in this chapter (i.e., unconstrained nonlinear optimization problems) can be improved by adding a local optimizer. This is evident in the results reported in Ugray, et al. (2001), where OCL is combined with an implementation of a generalized reduced gradient (GRG) procedure.

5.   CONCLUSIONS

In this chapter, we discussed the notion of optimizing a complex system when a "black-box" evaluator can be used to estimate performance based on a set of input values. In many business and engineering problems, the black-box evaluator has the form of a simulation model capable of mapping inputs into outputs, where the inputs are values for a set of decision variables and the outputs include the objective function value. We also addressed the development and application of a function library that uses scatter search in the context of optimizing complex systems.

In addition to the C Callable Library, the OptQuest optimizer is also available as a Visual Basic or Fortran callable library. Recent versions also include object-oriented versions written in C++ and Java. The reader may obtain updated versions of OptQuest directly from OptTek Systems, Inc. (www.opttek.com).

6. APPENDIX

This appendix contains the description of the set of test functions in Table 9-1. The description consists of the objective function, parameter values and the bounds for each variable. Note that all the problems in this appendix are minimization problems.

1. Branin

$$f(x) = \left(x_2 - \left(\frac{5}{4\pi^2} \right) x_1^2 + \left(\frac{5}{\pi} \right) x_1 - 6 \right)^2 + 10 \left(1 - \frac{1}{8\pi} \right) \cos(x_i) + 10$$

$-5 \le x_1, x_2 \le 15$

---

2. B2

$$f(x) = x_1^2 + 2x_2^2 - 0.3\cos(3\pi x_1) - 0.4\cos(4\pi x_2) + 0.7$$

$-50 \le x_i \le 100$ for $i=1, 2$

---

3. Easom

$$f(x) = -\cos(x_1)\cos(x_2)\exp\left(-\left((x_1-\pi)^2 + (x_2-\pi)^2\right)\right)$$

$-100 \le x_i \le 100$ for $i=1, 2$

4. Goldstein and Price

$$f(x) = \left(1 + (x_1 + x_2 + 1)^2\left(19 - 14x_1 + 3x_1^2 - 14x_2 + 6x_1x_2 + 3x_2^2\right)\right)$$
$$\left(30 + (2x_1 - 3x_2)^2\left(18 - 32x_1 + 12x_1^2 + 48x_2 - 36x_1x_2 + 27x_2^2\right)\right)$$

$-2 \le x_i \le 2$ for $i=1, 2$

5. Shubert

$$f(x) = \left(\sum_{j=1}^{5} j\cos((j+1)x_1 + j)\right)\left(\sum_{j=1}^{5} j\cos((j+1)x_2 + j)\right)$$

$-10 \le x_i \le 10$ for $i=1, 2$

6. Beale

$$f(x) = (1.5 - x_1 + x_1x_2)^2 + (2.25 - x_1 + x_1x_2^2)^2 + (2.625 - x_1 + x_1x_2^3)^2$$

$-4.5 \le x_1, x_2 \le 4.5$

7. Booth

$$f(x) = (x_1 + 2x_2 - 7)^2 + (2x_1 + x_2 - 5)^2$$

$-10 \leq x_1, x_2 \leq 10$

8. Matyas

$$f(x) = 0.26(x_1^2 + x_2^2) - 0.48x_1x_2$$

$-5 \leq x_1, x_2 \leq 10$

9. SixHumpCamelBack

$$f(x) = 4x_1^2 - 2.1x_1^4 + \frac{1}{3}x_1^6 + x_1x_2 - 4x_2^2 + 4x_2^4$$

$-5 \leq x_1, x_2 \leq 5$

10, 23. Schwefel(*n*)

$$f(x) = 418.9829n + \sum_{i=1}^{n}\left(-x_i \sin\sqrt{|x_i|}\right)$$

$-500 \leq x_i \leq 500$ for $i=1,..., n$

11, 29, 34. Rosenbrock(n)

$$f(x) = \sum_{i=1}^{n/2} 100(x_{2i} - x_{2i-1}^2)^2 + (1 - x_{2i-1})^2$$

$-10 \leq x_i \leq 10$ for $i=1,..., n$.

12, 30, 35. Zakharov(n)

$$f(x) = \sum_{j=1}^{n} x_j^2 + \left(\sum_{j=1}^{n} 0.5 jx_j \right)^2 + \left(\sum_{j=1}^{n} 0.5 jx_j \right)^4$$

$-5 \leq x_i \leq 10$ for $i=1, ..., n$

13. De Joung

$$f(x) = x_1^2 + x_2^2 + x_3^2$$

$-2.56 \leq x_i \leq 5.12$ for $i=1, 2, 3$

14. Hartmann(3,4)

$$f(x) = -\sum_{i=1}^{4} c_i \exp\left(-\sum_{j=1}^{3} a_{ij}(x_j - p_{ij})^2 \right)$$

$0 \leq x_i \leq 1$ for $i=1, 2, 3$

Table 9-2. Hartmann coefficients

i		a_{ij}		c_i		p_{ij}	
1	3.0	10.0	30.0	1.0	0.3689	0.1170	0.2673
2	0.1	10.0	35.0	1.2	0.4699	0.4387	0.7470
3	3.0	10.0	30.0	3.0	0.1091	0.8732	0.5547
4	0.1	10.0	35.0	3.2	0.0381	0.5743	0.8828

15. Colville

$$f(x) = 100(x_2 - x_1^2)^2 + (1 - x_1)^2 + 90(x_4 - x_3^2)^2 + (1 - x_3)^2 +$$
$$10.1\big((x_2 - 1)^2 + (x_4 - 1)^2\big) + 19.8(x_2 - 1)(x_4 - 1)$$

$-10 \le x_i \le 10$ for $i=1,...,4$.

16-18. Shekel(n)

$$f(x) = -\sum_{i=1}^{n} \big((x - a_i)^T (x - a_i) + c_i\big)^{-1} ;$$
$$x = (x_1, x_2, x_3, x_4)^T ; a_i = (a_i^1, a_i^2, a_i^3, a_i^4)^T$$

$0 \le x_i \le 10$ for $i=1, 2, 3, 4$

Table 9-3. Shekel coefficients

i		a_i^T			c_i
1	4.0	4.0	4.0	4.0	0.1
2	1.0	1.0	1.0	1.0	0.2
3	8.0	8.0	8.0	8.0	0.2
4	6.0	6.0	6.0	6.0	0.4
5	3.0	7.0	3.0	7.0	0.4
6	2.0	9.0	2.0	9.0	0.6
7	5.0	5.0	3.0	3.0	0.3
8	8.0	1.0	8.0	1.0	0.7
9	6.0	2.0	6.0	2.0	0.5
10	7.0	3.6	7.0	3.6	0.5

19. Perm(n, β)

$$f(x) = \sum_{k=1}^{n} \left(\sum_{i=1}^{n} \left(i^k + \beta \right) \left(\left(\frac{x_i}{i} \right)^k - 1 \right) \right)^2$$

$-n \leq x_i \leq n$ for $i=1,..., n$.

20. Perm0(n, β)

$$f(x) = \sum_{k=1}^{n} \left(\sum_{i=1}^{n} \left(i + \beta \right) \left(x_i^k - \left(\frac{1}{i} \right)^k \right) \right)^2$$

$-n \leq x_i \leq n$ for $i=1,..., n$.

21. PowerSum($b_1,...,b_n$)

$$f(x) = \sum_{k=1}^{n} \left(\left(\sum_{i=1}^{n} x_i^k \right) - b_k \right)^2$$

$0 \leq x_i \leq n$ for $i=1,..., n$.

22. Hartmann(6,4)

$$f(x) = -\sum_{i=1}^{4} c_i \exp\left(-\sum_{j=1}^{6} a_{ij} \left(x_j - p_{ij} \right)^2 \right)$$

$0 \leq x_i \leq 1$ for $i=1, ..., 6$

Table 9-4. Hartmann coefficients

i				a_{ij}			c_i
1	10.0	3.0	17.0	3.5	1.7	8.0	1.0
2	0.05	10.0	17.0	0.10	8.0	14.0	1.2
3	3.0	3.5	1.7	10.0	17.0	8.0	3.0
4	17.0	8.0	0.05	10.0	0.1	14.0	3.2
i				p_{ij}			
1	0.1312	0.1696	0.5569	0.0124	0.8283	0.5886	
2	0.2329	0.4135	0.8307	0.3736	0.1004	0.9991	
3	0.2348	0.1451	0.3522	0.2883	0.3047	0.6650	
4	0.4047	0.8828	0.8732	0.5743	0.1091	0.0381	

24-25. Trid(n)

$$f(x) = \left(\sum_{i=1}^{n} (x_i - 1)^2 \right) - \sum_{i=2}^{n} x_i x_j$$

$-n^2 \le x_i \le n^2$ for $i=1,...,n$.

26, 31. Rastrigin(n)

$$f(x) = 10n + \sum_{i=1}^{n} \left(x_i^2 - 10 \cos(2\pi x_i) \right)$$

$-2.56 \le x_i \le 5.12$ for $i=1,...,n$

27, 32. Griewank(n)

$$f(x) = \sum_{i=1}^{n} \frac{x_i^2}{4000} - \prod_{i=1}^{n} \cos\left(\frac{x_i}{\sqrt{i}} \right) + 1$$

$-300 \le x_i \le 600$ for $i=1,...,n$

28, 33. Sum Squares (*n*)

$$f(x) = \sum_{i=1}^{n} ix_i^2$$

$-5 \le x_i \le 10$ for $i=1,..., n$

36. Powell(*n*)

$$f(x) = \sum_{j=1}^{n/4}(x_{4j-3}+10x_{4j-2})^2 + 5(x_{4j-1}-x_{4j})^2 + (x_{4j-2}-2x_{4j-1})^4 + 10(x_{4j-3}-x_{4j})^4$$

$-4 \le x_i \le 5$ for $i=1,..., n$

37. Dixon and Price(*n*)

$$f(x) = \sum_{i=1}^{n} i(2x_i^2 - x_{i-1})^2 + (x_1 - 1)^2$$

$-10 \le x_i \le 10$ for $i=1,..., n$

38. Levy(n)

$$f(x) = \sin^2(\pi y_1) + \sum_{i=1}^{n-1}(y_i - 1)^2(1 + 10\sin^2(\pi y_i + 1)) + (y_n - 1)^2(1 + \sin^2(2\pi x_n))$$

$$y_i = 1 + \frac{x_i - 1}{4} \qquad\qquad \text{for } i=1,\dots,n$$

$$-10 \le x_i \le 10 \qquad\qquad \text{for } i=1,\dots,n$$

39. Sphere(n)

$$f(x) = \sum_{i=1}^{n} x_i^2$$

$$-2.56 \le x_i \le 5.12 \qquad\qquad \text{for } i=1,\dots,n$$

40. Ackley(n)

$$f(x) = 20 + e - 20e^{-0.2\sqrt{\frac{1}{n}\sum_{i=1}^{n}x_i^2}} - e^{\frac{1}{n}\sum_{i=1}^{n}\cos(2\pi x_i)}$$

$$-15 \le x_i \le 30 \qquad\qquad \text{for } i=1,\dots,n$$

Chapter 10

EXPERIENCES AND FUTURE DIRECTIONS

I think this is the beginning of a beautiful friendship.

Humphrey Bogart in Casablanca (1942)

The development and implementation of metaheuristic procedures usually entails a fair amount of experimentation and reliance on past experiences. Metaheuristic methodology is based on principles and not necessarily on theory that can be spelled out with theorems and proofs. The operations research community has for quite some time accepted the notion of creating search methodologies under the umbrella of *modern optimization techniques*. Twenty five years ago, however, the same community of researchers looked at heuristics in a different light, as evident by this quote from Glover (1977):

"Algorithms are conceived in analytic purity in the high citadels of academic research, heuristics are midwifed by expediency in the dark corners of the practitioner's lair ... and are accorded lower status."

The acceptance of the so-called modern optimization techniques is mainly due to the important contributions that metaheuristic research has made in the area of solving practical optimization problems. In other words, for metaheuristics "the proof of the pudding is in the eating."[1] Consequently, our primary goal in this book has been to provide a practical "hands-on" experience by including tutorial chapters and computer code. Throughout the book, we have encouraged the reader to experiment with the accompanying computer codes and create improved designs because we believe that in this case "skill comes from practice." In this last chapter, we first summarize our experiences on the subject of implementing scatter

---

[1] Miguel de Cervantes Saavedra (1547–1616), *Don Quijote,* Part II, Chapter XXIV.

search in several problem settings. We then conclude with a discussion on possible directions for future scatter search research.

1.   EXPERIENCES AND FINDINGS

Because we have experimented with scatter search for a number of years, we believe that we have learned some valuable lessons that are the result of implementing ideas and engaging on extensive experimental testing. Although in one way or another we have already discussed most of these findings throughout the book, we find it helpful to summarize these key implementation guidelines in this final chapter. We classify our thirteen findings into five categories associated with the methods that make up a scatter search implementation.

1.1   Diversification Generation

The first set of findings relate to the method for generating diversification within scatter search. We assume that a scatter search implementation will use the Diversification Generation Method both for initializing the reference set and also for rebuilding the reference set during the search.

Finding 1: *Frequency memory is an important element for creating effective diversification generation methods.*

We consider that a Diversification Generation Method is effective if it is capable of generating solutions maintaining a balance between solution quality and diversity. Section 4.2 of Chapter 5 describes 10 diversification generators for a scatter search implementation tackling the linear ordering problem (LOP). Since solutions to the LOP can be represented as permutations, construction heuristics that consider one element at a time are suitable. The first six generators (labeled DG01-DG06) are based on GRASP constructions. Therefore, these methods use a greedy function to measure the attractiveness of selecting an element for a given position in the permutation. The selection is random among a short list of attractive choices. The seventh procedure (DG07) is a hybrid combination of the first six. The eighth procedure (DG08) simply constructs random permutations. The ninth (DG09) is an implementation of a systematic generator for permutations that seeks diversity without relying on randomization and without reference to the objective function value. The last procedure (DG10) constructs permutations using a penalized greedy function. The penalties are based on a frequency memory that keeps track of the number of times an element has occupied a position in the permutations previously generated. Experimentation with these methods, summarized in Figure 5-9 of Chapter 5, showed that randomness alone creates diversification but neglects quality. We can observe in Figure 5-9 that DG10 has the best balance between quality and diversity. The pure random method DG08, on the other hand, is able to generate highly diverse solutions (as measured by Δd) but at the expense of quality. The use of frequency memory and a guiding function provides the desired balance exhibited in DG10.

We have found out that even those diversification methods that include random elements benefit from the use of frequency memory. The scatter search implementation in Chapter 2 uses frequency memory and controlled randomization in the context of nonlinear optimization. The range of each variable is divided into 4 sub-ranges of equal size. Then, a solution is constructed in two steps. First a sub-range is randomly selected. The probability of selecting a sub-range is inversely proportional to its frequency count. Then a value is randomly generated within the selected sub-range. The number of times a sub-range has been chosen to generate a value for a given variable is accumulated in an array. Experimentation showed that this frequency-based generator was superior to alternative designs based on randomization without the use of frequency memory.

Finding 2: *Initializing the reference set from a large set of solutions generated with the Diversification Generation Method is effective.*

The original scatter search proposal suggests a dynamic procedure for constructing the initial reference set. Specifically, the Diversification Generation Method is applied (followed by the Improvement Method) until b distinct solutions are generated. However, this procedure does not fully exploit the benefits of a Diversification Generation Method that is based on frequency memory and on diversification scheme that attempts to maximize the minimum distance between the candidate solutions and the solutions already in the reference set. To take advantage of such a method, the scatter search implementations in this book build a large set of solutions first. We refer to this set as P and to its corresponding size as *PopSize*. The set is built by applying the Diversification Generation Method followed by the Improvement Method until *PopSize* solutions are generated. The reference set is built by choosing b solutions from P. Typically the size of the set P is such that $p \geq 10 * b$.

1.2   Improvement Method

Most improvement methods in the context of scatter search consist of a simple local search. One should keep in mind that the Improvement Method must be capable of starting its descent (in the case of minimization) from an infeasible solution. The Improvement Method introduced in Chapter 3 for the knapsack problems seeks feasibility before improvement when the starting solution is infeasible. Improvement methods are applied to trial solutions generated with either the Diversification Generation Method or the Combination Method. Two key issues related to improvement methods are the frequency of their application and the depth of the associated search.

Finding 3: *A selective application of the Improvement Method may be better than applying it to very trial solution.*

Ugray, et al. (2002) develop a solution procedure for large-scale constrained nonlinear optimization problems. The procedure is based on combining scatter search with a large-scale GRG optimizer. LSGRG uses solutions generated with scatter search as the initial points for finding local optima. The initial solution may or may not be feasible (with respect to linear or nonlinear constraints), but LSGRG is designed to initiate the search from either infeasible or feasible points. Because a single execution of LSGRG is computational expensive, this optimizer is not applied to every solution generated during the scatter search. In this context, the scatter search must be calibrated to generate diverse initial points for a LSGRG exploration in order to avoid visiting the same local optima repeatedly.

Mazzini (1998) found, in the context of scheduling jobs on parallel identical machines, that an Improvement Method consisting of searching for the best move from either all possible swaps or insertions was computationally costly. To reduce the complexity, the Improvement Method was limited to examining a reduced list of candidate moves. Even with this limitation, the computational effort was only reduced when the Improvement Method was selectively applied. In particular, the Improvement Method was applied every 10 iterations to a diverse set of solutions selected from the solutions generated since the last application of the method and a small set of elite solutions. (Mazzini's scatter search design follows the scatter search description in Section 1.2 of Chapter 1.) More sophisticated selection rules were also tried without significantly improving the results.

Finding 4: *The application of the Improvement Method to all trial solutions accelerates the convergence of the reference set.*

In some situations, such as the scatter search implementations in Campos, et al. (2001), Corberán, et al. (2002), and Laguna and Martí (2000) and (2001b), the Improvement Method is such that it can be feasibly applied to all trial solutions. Although this is a desirable feature from the point of view of finding high quality solutions quickly, the reference set might "prematurely converge" (where convergence here means reaching a state in which no new solutions are admitted to the set). This quick convergence is not a limitation and in fact it might actually be desirable when the allotted searching time is short (e.g., in the context of optimizing simulations). Reference set convergence can also be tackled by rebuilding the set as discussed in Finding 7 of this chapter.

In Campos, Laguna and Martí (2001) the Improvement Method is used to enhance every trial solution. Because of the premature convergence, a reference set size that is larger than customary was tried. In particular, two instances of scatter search were used for experimentation, one with the reference set size $b = 20$ and the population size $PopSize = 100$ and the other one with $b = 40$ and $PopSize = 200$. The procedure stops when no new solutions are admitted into the reference set and therefore the length of the search depends on the size of the reference set. This design resulted in superior outcomes when compared to one where a small ($b = 10$) reference set was rebuilt several times. The best solutions were found with $b = 40$ at the expense of additional computational time.

Finding 5: *Seek a balance between the computational time spent generating trial solutions and the time spent improving such solutions.*

This is one of the most difficult issues in the design of scatter search procedures. Most improvement methods are local search procedures that stop as soon as no more improving moves can be made, resulting in a local optimal solution. The selection rule for choosing an improving move typically varies between selecting the first improving move and selecting the most improving move. The best improving strategy is sometimes computationally too costly, because it calls for finding the absolute best move available in a neighborhood. For instance, if a swap neighborhood is used to search in a solution space defined by permutations, the complete neighborhood of a solution consists of $(n^2 - n)/2$ solutions. The neighborhood size may be reduced by limiting the number of swaps examined. For example, there are only n-1 swaps of immediately consecutive elements in a permutation.

A first improving strategy seeks to find an improving move in the neighborhood of the current solution. When an improving move is found, the move is executed and the search continues from the new current solution. A best improving mechanism is not guaranteed to outperform a first improving strategy. Figure 10-1 shows a situation in a hypothetical minimization problem where "best improving" makes two moves before finding a local optimum while "first improving" results in five improving moves and a better local optimal solution.

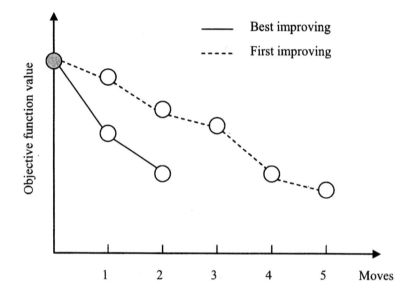

Figure 10-1. First improving vs. best improving

In the situation depicted in Figure 10-1, "first improving" is the better strategy from the point of view of final solution quality. In terms of computational time, the "winner" depends on how early the first improving moves are found, since the computational time to find the best improving move is fixed. In addition to considering these local optimization options, the scatter search designer must consider the use of metaheuristic optimizers as improving methods. Chapter 6 illustrates how a short-term memory tabu search can be integrated within the scatter search framework. A quick or fast annealing procedure may be considered as a possible metaheuristic Improvement Method.

1.3   Reference Set Update Method

Some of our most important findings relate to the way the reference set is initialized, updated, and rebuilt. We have experienced significant changes in the level of performance depending on the strategies used to update the reference set. These are our findings.

Finding 6: *Initialize the reference set with half of the set consisting of high quality solutions and the other half with diverse solutions.*

The right balance between intensification and diversification is a desirable feature of search procedures in general and scatter search in specific. The updating of the reference set can be used to achieve the appropriate balance in scatter search implementations. In Finding 2, we discussed the use of a set P to initialize the reference set. If the initialization of the reference set were made solely on the basis of solution quality, the set would be constructed selecting the best (in terms of quality) b solutions from P. If instead, we want to balance quality and diversity, we use the Reference Set Update Method implemented in our tutorial chapters, which can be summarized as follows:

1. Start with $P = \emptyset$. Use the Diversification Generation Method to construct a solution x. Apply the Improvement Method to x to obtain the enhanced solution x^*. If $x^* \notin P$ then, add x^* to P (i.e., $P = P \cup x^*$), otherwise, discard x^*. Repeat this step until $|P| = PopSize$.

2. Order the solutions in P according to their objective function value (where the best overall solution is first on the list).

3. Select the first b_1 solutions in P and add them to *RefSet*.

4. For each solution x in P-*RefSet*, calculate the minimum dissimilarity $d_{min}(x)$ to all solutions in *RefSet*.

5. Select solution x^* with the maximum dissimilarity $d_{min}(x^*)$ of all x in P-*RefSet*.

6. Add x^* to *RefSet* and delete x^* from P. If $|R| = b$ stop, otherwise go to 4.

This initialization procedure uses a max-min criterion to select diverse solutions to be added to the reference set. In particular, it selects the solution that maximizes the minimum value of a dissimilarity measure between the candidate solution and the solutions currently in the reference set. The dissimilarity measure d depends on the problem context. For example, the Euclidean distance can be used in problems whose solution representation is a vector of continuous variables, as shown in Chapter 2. Numerous experiments with several values of b_1 and b_2 indicate that an effective settings is to make $b_1 = b_2$.

Finding 7: *Rebuild the reference set with the Diversification Generation Method*

Some scatter search implementations do not include a rebuilding mechanism. That is, the search stops or is completely restarted when the reference set converges. As mentioned before, convergence refers to the situation where no new trail solutions are admitted to the reference set. In this situation we recommend rebuilding the reference set as described in Section 1.2.1 of Chapter 5. The procedure consists of keeping the first b_1 solutions in *RefSet* and deleting the remaining b_2 solutions. Then, the Diversification Generation Method is executed to add b_2 solutions to complete the reference set. Although there may be other ways of dealing with the convergence of the reference set, employing the Diversification Generation Method to replace the worst b_2 solutions in the reference set has yielded improved outcomes. As customary, the solutions are added to *RefSet* using a max-min criterion that encouraged diversity.

Finding 8: *Solution quality is more important than the diversity of the solutions when updating the reference set.*

The scatter search implementations in the tutorial chapters 2, 3 and 4 employ a Reference Set Update Method that focuses on the quality of the trial solutions being considered for admission to the reference set. If the quality of a trial solution exceeds the quality of the worst solution in the reference set, then the new trial solution replaces the worst solution in *RefSet*. In Section 1.2.2 of Chapter 5, we introduced the notion of multi-tier reference sets. The 2-tier design divides the reference set into two subsets: *RefSet*$_1$ consisting of the best (according to quality) b_1 solutions and *RefSet*$_2$ consisting of the b_2 most diverse solutions. The diversity of the solutions in *RefSet*$_2$ is measured with a max-min criterion and *RefSet* = *RefSet*$_1$ ∪ *RefSet*$_2$. This updating mechanism is helpful in settings where the search requires additional diversity. For instance, in Ugray et al. (2002) the Generalized Reduced Gradient procedure used as the Improvement Method is such that solutions that are in the same basin of attraction converge to the same local optimum. Therefore, it is important in such a context to generate trial solutions that are "far away" from each other.

In many scatter search implementations, however, the additional diversification is not needed and therefore a 2-tier reference set must be used with caution. If experimentation shows that the search lacks intensification, we recommend decreasing the size of *RefSet₂*. This does not contradict our Finding 6, because that finding refers to designs that are not coupled with a multi-tier reference set.

Finding 9: *Compare the merit of static versus a dynamic updating method.*

There are basically two ways of updating the reference set: statically and dynamically. The static update is the one that was implemented in the tutorial chapters. In this update, the Combination and Improvement methods are applied to all the subsets generated with the Subset Generation Method in order to create the set X of all new trial solutions. The new reference set then consists of the best b solutions from *RefSet* $\cup$ X when the *RefSet* is not divided into several tiers. This updating mechanism guarantees that a reference solution is used at least once by the Combination Method.

Alternatively, the reference set may be dynamically updated as described in Section 1.1 of Chapter 5. The dynamic update consists of testing trial solutions for admission to *RefSet* as these solutions are generated. The appeal of the dynamic updating is that at the beginning of the search, low quality solutions are immediately replaced with higher quality solutions and thus making the method more aggressive. Starting the search with a reduced reference set size and using a static update may result in a similar effect. Suppose that the desired reference set size is $b = 15$. To make the search more aggressive within a static updating method, the procedure could start with $b = 10$. The number of trial solutions generated with the initial reference set is reduced and the poor solutions in the reference set are replaced faster. After an initial phase, the size of the reference set can be increased when no new solutions are admitted to the set. This monotonic increase could stop after reaching the target size of 15.

1.4   Subset Generation Method

New trial solutions in scatter search are generated from the combination of two or more reference solutions. In the tutorial chapters of this book, we limited the scope to combinations of two reference solutions. Then in Section 2 of Chapter 5, computer code for combinations of more than two solutions was introduced. The procedure seeks to generate subsets that have useful properties, while avoiding the duplication of subsets previously generated.

Finding 10: *Combination of 2-element subsets seem to "do most of the work".*

This finding is related to the experiment by Campos, et al. (2001b) described in Section 2 of Chapter 5 and summarized in Figure 5-5. The experiment was designed to assess the contribution of the different types of combinations embedded in an implementation of scatter search for the linear ordering problem. The goal of the experiment was to determine whether the best solutions were generated from the combination of 2, 3, 4 or all b reference solutions. That is, the experiment attempted to identify how often, across a set of benchmark problems, the best solutions came from various combinations of k reference solutions. The experiment showed that at least 80% of the solutions that were admitted to the reference set came from combinations of 2-element subsets.

Our observations are not meant to discourage the implementation of combination methods that operate on more than 2 solutions. What was observed by Campos, et al. (2001) is that although 2-element subsets seem to do most of the work there is a contribution from the combination of larger subsets. Additional experimentation is necessary to determine whether Finding 10 holds in a more general sense. The order in which the subsets are generated and combined may be important. Campos, et al. (2001b) processed subsets types in increasing size, i.e., 2-element subsets were combined first, followed by 3-element subsets, and so on. Therefore, the percentage of time that the best solutions are generated with the combination of two solutions may change if the subset types were generated and combined in a different sequence.

1.5   Combination Method

As we have discussed throughout the book, the Combination Method is an element of scatter search that is context-dependent. Although it is possible to design "generic" combination procedures, as done in context-independent implementations and commercial applications, it is generally more effective to base the design on specific characteristics of the problem setting.

Finding 11: *The best solutions are often generated by the combination of other high quality solutions.*

This finding relates to the discussion in Section 3.1 of Chapter 5 about an experiment that attempts to identify the ranks of the reference solutions that generate the best solutions during the search. The experiment considers that the reference set is ordered with the solution in the first rank being the best overall solution. A $b \times b$ matrix *Source(i,j)* is created, where *Source(i,j)* counts the number of times a solution of rank j is a reference solution for the current solution for rank i. The experiment showed that solutions of rank 1 are more often generated by combinations that involve other solutions of rank 1. This finding prompts the consideration of a strategy that generates more trial solutions when combining the best solutions in the reference set than when combining the worst solutions. The following Combination Method illustrates this in the context of nonlinear optimization and is an instance of the general description provided at the end of Section 3.1 in Chapter 5.

Let x and y be two solutions in the reference set that are used to perform three types of combinations to generate a trial solution z:

C1: $z = x - d$
C2: $z = x + d$
C3: $z = y + d$

where $d = 0.5r(y - x)$ and r is a number between 0 and 1. The combination rules are:

- If both x and y rank among the first b_1 solutions in *RefSet*, then generate 4 solutions by applying C1 and C3 once and C2 twice.
- If either x or y ranks among the first b_1 solutions in *RefSet*, then generate 3 solutions by applying C1, C2 and C3 once.
- If neither x nor y ranks among the first b_1 solutions in *RefSet*, then generate 2 solutions by applying C2 once and either C1 or C3 once.

Note that when a combination is applied twice, different values for r must be selected.

Finding 12: *Combination methods that include some randomness can be effective.*

The combination mechanism illustrated in the previous finding contains a parameter r that determines the combination weight and that ranges between 0 and 1. This Combination Method is similar to the one suggested in Glover (1994b) and listed as the fourth element in the Scatter / Tabu Search Hybrid description in Section 1.2 of Chapter 1. The difference between the procedure above and the one suggested in Glover (1994b) is that Glover prescribes five values for the weight parameter w, making the method completely deterministic. By allowing r to be a random number between 0 and 1, the Combination Method may be applied to the same reference solutions resulting in different outcomes. This flexibility can be used for intensification purposes, where more solutions can be sampled from a promising line with the same mechanism.

Finding 13: *The use of multiple combination methods can be effective.*

Genetic algorithms (GA) have been particularly good at exploiting this finding. Most GA implementations use several "operators" to combine parents and generate new trial solutions. The rules to select from the set of available operators vary from one implementation to another. An analogous process can be implemented within the scatter search framework, as shown in Section 1.1.2 of Chapter 7.

Adaptive structured combination methods provide a mechanism for implementing multiple forms of combinations within a single framework. In this approach, each reference solution proposes votes for a particular attribute to be incorporated in the new trial solution. The voting scheme is adaptive because a new trial solution is constructed with the sequential addition of the attributes that have obtained the most votes. Variations of the voting scheme result in different ways of combining solutions (which in itself can be viewed as creating multiple combination methods or "operators"). Adaptive structured combinations have been implemented in scatter search procedures for production scheduling (Mazzini 1998; Laguna, et al. 2000), project scheduling (Valls, et al. 1998) and linear ordering problems (Campos, et al. 2001), among others.

2.   MULTIOBJECTIVE SCATTER SEARCH

A promising direction for future scatter search research is its application in multiobjective optimization. Evolutionary Algorithms are well-suited for tackling optimization problems with multiple objectives, as stated in Coello, Van Veldhuizen and Lamont (2002). The multiobjective optimization problem is the problem of finding values for a set of decision variables that satisfy a set of constraints and that optimize a vector function whose elements represent the objectives. These functions are a mathematical representation of performance criteria that are usually in conflict. Hence, the meaning of "optimizing" is finding a solution that results in objective function values that are acceptable to a decision maker. Section 2 of Chapter 8, summarizes a scatter search implementation that deals with a multiobjective optimization problem. Specifically, the problem of routing school buses is approached with scatter search, where the objective of minimizing the number of buses is simultaneously considered with the objective of minimizing the longest time a student would have to stay in the bus. Clearly, these are conflicting objectives since the shortest "riding time" is achieved by picking up each student with a different bus. At the other end of the spectrum the solution that uses a single bus for all the students results in the maximum riding time for the student that was picked up first.

The school bus problem has two objectives only and one of them can be easily enumerated. In other words, since the maximum number of buses in any solution is a fairly small number, the problem can be solved as a sequence of single-objective problems that attempt to minimize the maximum riding time for each value of the number of buses. Figure 10-2 depicts the solutions obtained with scatter search (SS) compared to the solutions found with a constructive heuristic (H1) in Corberán, et al. (2002). In Figure 10-2, the maximum route length is obtained for one problem where the number of buses is varied from 5 to 21.

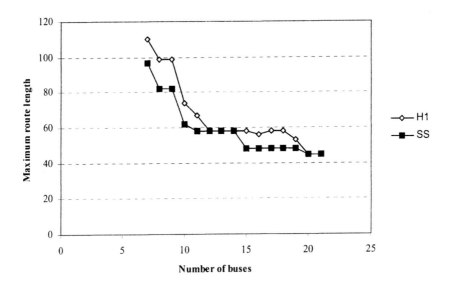

Figure 10-2. Bus routing with two objectives

In general, however, it is not always possible to solve multiobjective optimization problems as sequence of single-objective problems and therefore solutions methods in this area must be prepared to deal with more than one objective at a time. The notion of optimum in multiobjective optimization changes form finding a single global optimal value to finding a good compromise. In the bus routing example of Figure 10-2, the decision maker may choose from the sixteen alternatives available to her. Clearly, if the maximum length must be kept below one hour, then the solution with 11 buses is the best, since the same maximum length is associated with solutions that utilize 12, 13 and 14 buses. The trade-off in this problem is between service level, which is represented by the maximum route length, and operational cost, which is represented by the number of buses in the solution.

Pareto optimality is the most common notion of optimality in multiobjective optimization. Let x be a feasible solution to a multiobjective optimization problem with k objective functions $f_i(x)$ for $i = 1, ..., k$. Then, x^* is a Pareto optimal solution if for every feasible solution x either:

$$f_i(x^*) = f_i(x) \qquad \text{for } i = 1, ..., k$$

$$\text{or } f_i(x^*) < f_i(x) \qquad \text{for at least one } i$$

This definition considers that all the objective functions have been transformed to a minimization form. The interpretation of the definition is that if x^* is Pareto optimal then there exists no feasible solution x that would decrease the value of one of the functions without causing a simultaneous increase in at least one of the other functions. The notion of dominance (or dominated solutions) is used to define the *Pareto front* as the set of nondominated solutions in the decision variable space. The Pareto front is also known as the efficiency frontier, because Pareto optimal solutions are also referred to as efficient solutions.

Multiobjective problems can be classified in three categories according to the articulation of preferences (Coello, Van Veldhuizen and Lamont, 2002):

- *A priori* — This group of models include those approaches that assume the decision maker can articulate preferences (e.g., desired goals or ordering of the objectives) proper to the search. Goal programming and multiattribute utility theory fall within this category.

- *A posteriori* — These models do not require preference information prior to the search. The models associated with this category are among the oldest approaches to multiobjective optimization and include linear combination of weights and ε-constraint method.

- *Progressive* — These techniques include three stages: 1) finding a nondominated solution, 2) obtaining the reaction from the decision maker regarding the solution, and 3) modifying the preferences associated with each objective before repeating steps 1 and 2. PROTRADE (probabilistic trade-off development method) and STEM (STEP method) are in this category.

Most metaheuristic applications in the area of multiobjective optimization use models with a posteriori articulation of preferences. Among the techniques based on these models, the Pareto sampling technique commands the greatest attention in the world of multiobjective evolutionary approaches. To complete this section, we briefly discuss how scatter search could be adapted for the solution of the most common multiobjective models with a posteriori articulation of preferences.

2.1 Independent Sampling Technique

This technique consists of assigning a weight value to each objective function and executing the search procedure on the resulting single-objective problem. The weights are then uniformly varied and the search is performed again. The difference between this procedure and one in which the objective functions are simply aggregated, is that variability of the weights allows the generation of larger portions of the Pareto front during the search.

The adaptation of scatter search to perform an independent sampling of the Pareto front is straightforward. A scheme to vary the weights is designed and each run of scatter search is performed considering a single objective function that is a weighted combination of the original objective functions in the problem. The use of this approach is limited, because the number of scatter search runs required to obtain a reasonable picture of the Pareto front exponentially increase with the number of objective functions in the problem.

2.2 Criterion Selection Technique

The idea behind this technique is to partition a population (in genetic algorithms) or a reference set (in scatter search) in such a way that each objective function is properly represented in the entire set of solutions. The technique is simple and intuitive. It approximates the Pareto front in a single execution of the search procedure. The critical steps in a scatter search implementations would be:

1. *RefSet* is divided into k subsets *RefSet*(i) for $i = 1, .., k$.

2. After the application of the Diversification Generation Method or the Combination Method, admission to *RefSet*(i) is according to solution quality as measure by $f_i(x)$.

3. The Subset Generation Method is applied to *RefSet* followed by the Combination and Improvement methods. New trial solutions are tested

for membership in *RefSet* by checking for dominance in each of the k subsets.

The procedure should keep a separate set of solutions that are identified as belonging to the estimated Pareto front. Note that nondominance is limited to the current set of solutions because a dominated solution in the local sense is also dominated in the global sense but the opposite is not necessarily true.

2.3   Aggregation Selection Technique

This technique is similar to the independent sampling technique in that the search deals with a weighted combination of the objectives. The most common form of the aggregated objective function is the weighted linear combination:

$$f(x) = \sum_i w_i f_i(x)$$

The difference between aggregation selection and independent sampling is that in aggregation selection the weights are changed within the search while in independent selection the weights are static within the search but changed between searches. The aggregation selection technique can generate a set of solutions to approximate the Pareto front in a single run. The main disadvantage of this approach is that the weighted linear combination may miss members of the Pareto front even when randomly varying the associated weights.

2.4   Pareto Sampling

The basic idea of Pareto sampling techniques is to identify nondominated solutions from the current population of solutions. In genetic algorithms, the dominance criterion has been used to design several procedures based on Pareto ranking. For example, *Goldberg's Pareto ranking* finds nondominated solutions in the current populations to assign ranks before the population is evolved to the next generation (Goldberg, 1989). At a given iteration of a genetic algorithm, the set of nondominated solutions is identified and removed from the current population. The solutions that are removed first obtain the highest rank for the evolution step. The next set of nondominated solutions is identified from the solutions remaining in the current population and they are given the next highest rank. The process continues until the population is suitably ranked.

Fonseca and Fleming (1993) modified Goldberg's ranking procedure to make the rank of a solution dependent on the number of solutions in the current population that dominates it. For instance, the rank of a solution x may be given by $1+d(x)$, where $d(x)$ represents the number of solutions in the current population that dominate x. This procedure assigns a rank of 1 to all nondominated solutions and penalizes the dominated solutions according to the density of the corresponding trade-off region. This procedure is known in the literature as MOGA (multiobjective genetic algorithm).

There are several ways in which these ideas can be adapted in the context of scatter search. For example, the initial *RefSet* could be built by sequentially identifying nondominated solutions from P in the same way that Goldberg's ranking procedure operates. A static updating of the reference set could be used where all the new trial solutions are placed in a *Pool* (see e.g. Section 3 of Chapter 2). The reference set could then be updated by sequentially applying the nondominated criterion to the set of solutions that result from the union of the current reference set and *Pool*.

The equivalent of MOGA in scatter search is using the ranking function to assign a rank to the solutions in P and then select those with the highest rank for *RefSet*. Similarly, after an application of the Combination Method, *Pool* $\cup$ *RefSet* may be ranked and the reference set may be updated with the highest ranked solutions. Issues of diversity and the consideration of a multi-tier *RefSet* should be explored to keep the search from focusing on a given part of the Pareto front while neglecting the rest.

3. MAXIMUM DIVERSITY PROBLEM

The success of the scatter search method is mainly due to a good balance between quality and diversity in the reference set. In this section we propose some new research directions to monitor and possibly increasing the diversity of *RefSet*.

In the tutorial chapters we introduced a max-min method to select a set of diverse solutions from P to be added to *RefSet*. This mechanism was used for the initial construction as well as the rebuilding of the reference set. Specifically, the solution in P with the maximum minimum distance to the solutions in currently in *RefSet* was chosen to be added to the reference set. The distance measure for the application of the max-min criterion was customized to each situation. Our procedure may be viewed as a heuristic mechanism for selecting a set of solutions with maximum diversity. This problem has been studied in the literature and several procedures have been developed to find heuristic and exact solutions for it.

The problem of selecting a subset of maximum diversity from a given set of elements is known as the *maximum diversity problem*. This problem also arises in a wide range of real-world settings, and it has been the subject of several previous studies (Kuo et al., 1993; Glover, Kuo and Dhir, 1998). As an avenue for future research, it may be advantageous to exploit the knowledge associated with solving this problem to design different mechanisms for selecting a maximum diverse set of solutions in P to be added to *RefSet* in a scatter search procedure. The problem can be formally stated as follows.

Let $S = \{s_i : i \in N\}$ be a set of elements where $N = \{1, 2,.., n\}$ and each element s_i can be represented as a vector $s_i = (s_{i1}, s_{i2}, ..,s_{ir})$. Let d_{ij} be the distance between elements s_i and s_j, and let $m < n$ be the desired size of the diverse set. Then, the problem consists of selecting m elements in S in order to maximize the sum of the distances between the selected elements:

$$Maximize \sum_{i<j} d_{ij} x_i x_j$$

subject to:

$$\sum_{i=1}^{n} x_i = m$$

$$x_i \in \{0,1\} \quad i = 1,..,n$$

where $x_i = 1$ indicates that element s_i has been selected. This formulation appears in Kuo, et al. (1993) but it is not solved directly. The formulation is transformed to equivalent linear integer programs that offer greater computational efficiency. One of these formulations was used as the basis for showing that the maximum diversity problem is NP-hard.

In Glover, et al. (1998) four different heuristics are proposed for this problem. Since different versions of this problem include additional constraints, the objective is to design heuristics whose basic moves for transitioning from one solution to another are both simple and flexible, allowing these moves to be adapted to multiple settings. Moves that are especially attractive in this context are *constructive* and *destructive* moves that drive the search to approach and cross feasibility boundaries from different directions (see strategic oscillation in Section 2.2 of Chapter 7). Such moves are also highly natural in the maximum diversity problem, where the goal is to determine an optimal composition for a set of selected elements.

In Glover, et al. (1998) two constructive and two destructive heuristics are proposed. The first constructive and destructive methods are based on the concept of the center of gravity of a set. The center, $s\_center(X)$, of a set of elements $X = \{s_i : i \in I\}$ is define as:

$$s\_center(X) = \frac{\sum_{i \in I} s_i}{|X|}$$

Figure 10-3 presents an outline of the first constructive heuristic, where S is the set of elements and Sel denotes the set of selected elements. The first destructive heuristic is shown in Figure 10-4.

---

1. $Sel = \varnothing$
2.   Compute $s_c = s\_center(S)$
   while ($|Sel| < m$)
      3.   Let $i^* / d(s_{i*}, s_c) = \max_{s_i \in S}\{d(s_i, s_c)\}$
      4. $Sel = Sel \cup \{s_{i*}\}$
      5. $S = S - \{s_{i*}\}$
      6. $s_c = s\_center(Sel)$
   end while

Figure 10-3. Constructive heuristic C_1

---

1. $Sel = S$
2.   Compute $s_c = s\_center(S)$
   while ($|Sel| > m$)
      3.   Let $i^* / d(s_{i*}, s_c) = \min_{s_i \in S}\{d(s_i, s_c)\}$
      4. $Sel = Sel - \{s_{i*}\}$
      5. $s_c = s\_center(Sel)$
   end while

Figure 10-4. Destructive heuristic D_1

The second constructive and destructive heuristics are variations of the first ones, where instead of constructing a center of the set, the distance between an element s_i and a set $X = \{s_j : j \in I\}$ is defined as follows:

$$d(s_i, X) = \sum_{j \in I} d(s_i, s_j)$$

The heuristics in Figures 10-3 and 10-4 may be adapted to the problem of selecting b_2 solutions in P to be added to *RefSet*. In heuristic C_1, *RefSet* takes the place of *Sel*, P takes the place of S, and the value of m is replaced with b. The initialization step 1 is modified to include the b_1 high-quality solutions already in *RefSet*. Step 2 is also modified to calculate the center of gravity of *Sel*, $s_c = s\_center(Sel)$. That is, this step calculates the center of gravity of the b_1 solutions currently in *RefSet*. When the construction finishes, *Sel* contains the b solutions in *RefSet*.

In the D_1 heuristic of Figure 10-4, we consider that *Sel* is initially equal to P in step 1. In step 2, we compute s_c as the center of the b_1 high-quality solutions already in *RefSet*. The value of m is replaced with b_2 in the while loop and when the procedure finishes, the solutions in *Sel* are added to *RefSet*.

The heuristics based on distance measures instead of centers of gravity can also be adapted in the context of selecting diverse solutions to be added to a reference set. Experimentation with all of these procedures is necessary to determine the merit of using an approximate solution to the maximum diversity problem instead of the min-max criterion that we have employed throughout the book to select diverse solutions for the reference set.

4.   IMPLICATIONS FOR FUTURE DEVELOPMENTS

The focus and emphasis of scatter search and the path relinking approach described in Chapter 7 have a number of specific implications for the goal of designing improved optimization procedures. To understand these implications, it is useful to consider certain contrasts between the highly exploitable meaning of "solution combination" provided by path relinking and the rather amorphous concept of "crossover" used in genetic algorithms. Originally, GAs were founded on precise notions of crossover, using definitions based on binary strings and motivated by analogies with genetics. Although there are still many GA researchers who favor the types of crossover models originally proposed with genetic algorithms—since these give rise to the theorems that have helped to popularize GAs—there are also many who have largely abandoned these ideas and who have sought, on a case-by-case basis, to replace them with something different. The well-defined earlier notions of crossover have not been abandoned without a price. The literature is rife with examples where a new problem (or a new variant of an old one) has compelled the search for an appropriate "crossover" to begin anew.[2]

As a result of this lack of an organizing principle, many less-than-suitable modes of combination have been produced, some eventually replacing others, without a clear basis for taking advantage of context — in contrast to the strong context-exploiting emphasis embodied in the concept of search neighborhoods. The difficulty of devising a unifying basis for understanding or exploiting context in GAs was inherited from its original theme, which had the goal of making GAs *context free.*

A few of the more conspicuous features of "genetic crossover" and path relinking that embody such contrasts appear in Table 10-1.

Table 10-1. Comparison between GA and PR features

Genetic crossover features	Contrasting path relinking features
Contains no integrated framework	Embodies a unifying "path combination" principle
Each new "crossover" is separate, with no guidance for the next	Each implementation of path relinking derives from a common foundation
No basis exists to systematically exploit context	Context inheres in neighborhood structures and is directly exploitable by them
Advances are piecemeal, without clear sources of potential for transfer	Advances in neighborhood search foster advances in path relinking (and reciprocally)
There is no design plan that is subject to analysis or improvement	A cohesive framework exists for developing progressively improved methods

The differences identified in Table 10-1 have important consequences for research to yield improved methods. Specific areas of research for developing improved solution strategies that emerge directly from the path relinking orientation are catalogued as follows:

1. Connections and complementarities between neighborhoods for search methods and neighborhoods for path relinking
2. Rules for generating paths to different depths and thresholds of quality
3. Strategies for generating multiple paths between and beyond reference solutions (with parallel processing applications)
4. Path interpolations and extrapolations that are effective for intensification and diversification goals
5. Strategies for clustering and anti-clustering, to generate candidate sets of solutions to be combined
6. Rules for multi-parent compositions

---

[2] The disadvantage of lacking a clear and unified model for combining solutions has had its compensations for academic researchers, since each new application creates an opportunity to publish another form of crossover! The resulting abundance of papers has done nothing to tarnish the image of a dynamic and prospering field.

7. Isolating and assembling solution components by means of constructive linking and vocabulary building

These research opportunities carry with them an emphasis on producing systematic and strategically designed rules, rather than following the policy of relegating decisions to random choices, as often is fashionable in evolutionary methods. The strategic orientation underlying path relinking is motivated by connections with the tabu search setting where the path relinking ideas were first proposed, and invites the use of adaptive memory structures in determining the strategies produced. The learning approach called target analysis (Glover and Laguna, 1997) gives a particularly useful basis for pursuing such research.

REFERENCES

Aggarwal, C. C., J. B. Orlin and R. P. Tai (1997) "Optimized Crossover for the Independent Set Problem," *Operations Research*, vol. 45, pp. 226-234.

Aiex, R.M., M.G.C. Resende, P.M. Pardalos and G. Toraldo (2000) "GRASP with Path Relinking for the Three-index Assignment Problem," Technical Report.

Aiex, R.M., M.G.C. Resende and C.C. Ribeiro (2002) "Probability Distribution of Solution Time in GRASP: An Experimental Investigation," *Journal of Heuristics,* vol. 8, pp. 343-373.

Atan, T. and N. Secomandi (1999) "A Rollout-Based Application of the Scatter Search/Path Relinking Template to the Multi-Vehicle Routing Problem with Stochastic Demands and Restocking," Technical Report, PROS Revenue Management Inc., Houston, TX.

Balas, E. and W. Niehaus (1998) "Optimized Crossover-Based Genetic Algorithms for the Maximum Cardinality and Maximum Weight Clique Problems," *Journal of Heuristics*, vol. 4, no. 2, pp. 107-124.

Barnes, J. W. and L. K. Vanston (1981) "Scheduling Jobs with Linear Delay Penalties and Sequence Dependent Setup Costs," *Operations Research*, vol. 29, no. 1.

Becker, O. (1967) "Das Helmstädtersche Reihenfolgeproblem — die Effizienz verschiedener Näherungsverfahren," in *Computer Uses in the Social Sciences*, Berichteiner Working Conference, Wien.

Bulut, G. (2001) "Robust Multi-Scenario Optimization of an Air Expeditionary Force Structure Applying Scatter Search to the Combat Forces Assessment Model," M.S. Thesis, AFIT/GOR/ENS/01M-05, Air Force Institute of Technology.

Campos, V., M. Laguna and R. Martí (1999) "Scatter Search for the Linear Ordering Problem," in *New Ideas in Optimization*, Corne, Dorigo and Glover (Eds.), McGraw-Hill, pp. 331-341.

Campos, V., F. Glover, M. Laguna and R. Martí (2001) "An Experimental Evaluation of a Scatter Search for the Linear Ordering Problem," *Journal of Global Optimization*, vol. 21, no. 4, pp. 397-414.

Campos, V., M. Laguna and R. Martí (2001) "Context-Independent Scatter and Tabu Search for Permutation Problems," University of Colorado at Boulder, http://leeds.colorado.edu/faculty/laguna/articles/sstsperm.html.

Cavique, L., C. Rego and I. Themido (2001) "A Scatter Search Algorithm for the Maximum Clique Problem," in *Essays and Surveys in Metaheuristics*, C.C. Ribeiro and P. Hansen (Eds.), Kluwer Academic Publishers.

Canuto, S.A. (2000) "Local Search for the Prize-collecting Steiner Tree Problem," (in Portuguese), M.S. Thesis, Department of Computer Science, Catholic University of Rio de Janeiro.

Canuto, S. A., M. G. C. Resende and C. C. Ribeiro (2001) "Local Search with Perturbations for the Prize-Collecting Steiner Tree Problem in Graphs," *Networks*, vol. 38, pp. 50-58.

Chanas, S. and P. Kobylanski (1996) "A New Heuristic Algorithm Solving the Linear Ordering Problem," *Computational Optimization and Applications*, vol. 6, pp. 191-205.

Coello, C. A., D. A. Van Veldhuizen and G. B. Lamont (2002) *Evolutionary Algorithms for Solving Multi-Objective Problems*, Kluwer Academic / Plenum Publishers, New York.

Consiglio, A. and S. A. Zenios (1999) "Designing Portfolios of Financial Products via Integrated Simulation and Optimization Models," *Operations Research*, vol. 47, pp. 195-208.

Corberán, A., E. Fernández, M. Laguna and R. Martí (2002) "Heuristic Solutions to the Problem of Routing School Buses with Multiple Objectives," *Journal of the Operational Research Society*, vol. 53, no. 4, pp. 427-435.

Cung, V-D., T. Mautor, P. Michelon and A. Tavares (1996) "Scatter Search for the Quadratic Assignment Problem," Technical Report #1996/37, Laboratoire PRiSM, http://www.prism.uvsq.fr/rapports/1996/docuemtn_1996_37.ps.

Cung, V.-D., T. Mautor, P. Michelon and A. Tavares (1997) "A Scatter Search Based Approach for the Quadratic Assignment Problem," in *Proceedings of IEEE-ICEC-EPS'97, IEEE International Conference on Evolutionary Computation and Evolutionary Programming Conference*, T. Bäck, Z. Michalewicz and X. Yao (Eds.), pp. 165–170.

Cung, V.D., S.L. Martins, C.C. Ribeiro and C. Roucairol (2001) "Strategies for the Parallel Implementation of Metaheuristics," in *Essays and Surveys in Metaheuristics*, C.C. Ribeiro and P. Hansen (Eds.), Kluwer Academic Publishers, pp. 263-308.

Delgado, C., M. Laguna and J. Pacheco (2002) "Minimizing Labor Requirements in a Periodic Vehicle Loading Problem," Technical Report, University of Burgos, Spain.

Dorne, R. and J.K. Hao (1998) "Tabu Search for Graph Coloring, T-Colorings and Set T-Colorings," in *Meta-heuristics: Advances and Trends in Local Search Paradigms for Optimization*, S. Voss, S. Martello, I.H. Osman and C. Roucairol (Eds.), Kluwer Academic Publishers, pp. 77-92.

Dueck, G. H. and J. Jeffs (1995) "A Heuristic Bandwidth Reduction Algorithm," *J. of Combinatorial Math. and Comp.*, vol. 18, pp. 97-108.

Eades, P. and D. Kelly (1986) "Heuristics for Drawing 2-Layered Networks," *Ars Combinatoria*, vol. 21, pp. 89-98.

Everett, H. (1963) "Generalized Lagrangean Multiplier Method for Solving Problems of Optimal Allocation of Resources," *Operations Research*, vol. 11, pp. 399-417.

Festa, P. and M. G. C. Resende (2001) "GRASP: An Annotated Bibliography," in *Essays and Surveys in Metaheuristics*, C. C. Ribeiro and P. Hansen (Eds.), Kluwer Academic Publishers, pp. 325-367.

Fleurent, C. and F. Glover (1997) "Improved Constructive Multistart Strategies for the Quadratic Assignment Problem Using Adaptive Memory," Technical Report, University of Colorado.

Fleurent, C., F. Glover, P. Michelon and Z. Valli (1996) "A Scatter Search Approach for Unconstrained Continuous Optimization," in *Proceedings of the 1996 IEEE International Conference on Evolutionary Computation*, pp. 643-648.

Fogel, D. B. (1998) "An Introduction to Evolutionary Computation," in *Evolutionary Computation: The Fossil Record*, D. B. Fogel (Ed.), IEEE Press, pp. 1 - 2.

Fonseca, C. M. and P. J. Fleming (1993) "Genetic Algorithms for Multiobjective Optimization: Formulation, Discussion and Generalization," in *Proceedings of the Fifth International Conference on Genetic Algorithms*, S. Forrest (Ed.), San Mateo California, pp. 416-423.

Funabiki, N. and T. Higashino (2000) "A Minimal-State Processing Search Algorithm for Graph Colorings Problems," *IEICE Transactions on Fundamentals*, vol. E83, no. A7, pp. 1420-1430.

García-López, F., B. Melián-Batista, J. A. Moreno-Pérez and J. M. Moreno-Vega (2002a) "Parallelization of the Scatter Search," Technical Report, Dpto. de E.I.O. y Computación, Universidad de La Laguna.

García-López, F., B. Melián-Batista, J. A. Moreno-Pérez and J. M. Moreno-Vega (2002b) "The Parallel Variable Neighborhood Search for the p-Median Problem," *Journal of Heuristics*, vol. 8, pp. 375-388.

Galinier, P. and J.K. Hao (1999) "Hybrid Evolutionary Algorithms for Graph Coloring", *Journal of Combinatorial Optimization*, vol. 3, no. 4, pp. 379-397.

Glover, F. (1965) "A Multiphase-Dual Algorithm for the Zero-One Integer Programming Problem," *Operations Research*, vol. 13, pp. 879-919.

Glover, F. (1977) "Heuristics for Integer Programming Using Surrogate Constraints," *Decision Sciences*, vol. 8, pp. 156-166.

Glover, F. (1992) "Ejection Chains, Reference Structures and Alternating Path Methods for Traveling Salesman Problems," Technical Report, University of Colorado. Shortened version published in (1996) *Discrete Applied Mathematics*, vol. 65, pp. 223-253.

Glover, F. (1994a) "Genetic Algorithms and Scatter Search: Unsuspected Potentials," *Statistics and Computing*, vol. 4, pp. 131-140.

Glover, F. (1994b) "Tabu Search for Nonlinear and Parametric Optimization (with Links to Genetic Algorithms)," *Discrete Applied Mathematics*, vol. 49, pp. 231-255.

Glover, F. (1995) "Scatter Search and Star Paths: Beyond the Genetic Metaphor," *OR Spektrum*, vol. 17, pp. 125-137.

Glover, F. (1998) "A Template for Scatter Search and Path Relinking," in *Artificial Evolution, Lecture Notes in Computer Science 1363*, J.-K. Hao, E. Lutton, E. Ronald , M. Schoenauer and D. Snyers (Eds.), Springer, pp. 13-54.

Glover, F. (1999) "Scatter Search and Path Relinking," in *New Methods in Optimization*, D. Corne, M. Dorigo and F. Glover (Eds.), McGraw-Hill, London, pp. 297-316.

Glover, F. and G. Kochenberger (1996) "Critical Event Tabu Search for Multidimensional Knapsack Problems," in *Meta-heuristics: Theory and Applications*, I. H. Osman and J. P. Kelly (Eds.), Kluwer Academic Publishers, Boston, pp. 407-427.

Glover, F., C.C. Kuo and K.S. Dhir (1998) "Heuristic Algorithms for the Maximum Diversity Problem," *Journal of Information & Optimization Sciences*, vol. 19, pp. 109-132.

Glover, F. and M. Laguna (1997) *Tabu Search,* Kluwer Academic Publishers, Boston.

Glover, F., M. Laguna and R. Martí (2000) "Fundamentals of Scatter Search and Path Relinking," *Control and Cybernetics*, vol. 39, no. 3, pp. 653-684.

Glover, F., A. Løkketangen and D. L. Woodruff. (2000) "Scatter Search to Generate Diverse MIP Solutions," in *OR Computing Tools for Modeling, Optimization and Simulation: Interfaces in Computer Science and Operations Research*, M. Laguna and J.L. González-Velarde (Eds.), Kluwer Academic Publishers, pp. 299-317.

Ghamlouche, I., T.G. Crainic and M. Gendreau (2001) "Cycle-based Neighbourhoods for Fixed-Charge Capacitated Multicommodity Network Design," Technical Report CRT-2001-01, Centre de recherche sur les transports, Université de Montréal.

Ghamlouche, I., T.G. Crainic and M. Gendreau (2002) "Path Relinking, Cycle-Based Neighbourhoods and Capacitated Multicommodity Network Design," Technical Report CRT-2002-01, Centre de recherche sur les transports, Université de Montréal.

Goldberg, D. E. (1989) *Genetic Algorithms in Search, Optimization and Machine Learning,* Addison-Wesley Publishing Company, Reading, Massachusetts.

González-Velarde, J.L. and M. Laguna (2001) "Tabu Search with Simple Ejection Chains for Coloring Graphs," *Annals of Operations Research,* forthcoming.

Grant, R. S. (1998) "Evaluating the Use of Genetic Algorithms, Evolution Strategies, and Scatter Search for Performing Simulation Optimization," Ph.D. Dissertation, Mississippi State University.

Grabowski, J. and M. Wodecki (2001) "A New Very Fast Tabu Search Algorithm for the Job Shop Problem," Technical Report 21/2001, Instytut Cybernetyki Techncznej Politechniki Wroclawskiej, Wroclaw.

Grotschel, M., M. Junger and G. Reinelt (1984) "A Cutting Plane Algorithm for the Linear Ordering Problem," *Operations Research,* vol. 32, no. 6, pp. 1195-1220.

Greistorfer, P. (1995) "Computational Experiments with Heuristics for a Capacitated Arc Routing Problem," in *Operations Research Proceedings 1994,* U. Derigs, A. Bachem and A. Drexl, (Eds.), Springer-Verlag, Berlin, pp. 185-190.

Greistorfer, P. (1999) "Tabu and Scatter Search Combined for Arc Routing," presented at MIC'99 – 3rd Metaheuristics International Conference, Brazil, July 19-22, 1999.

Greistorfer, P. (2001a) "A Tabu Scatter Search Metaheuristic for the Arc Routing Problem," *Computers & Industrial Engineering,* forthcoming.

Greistorfer, P. (2001b) "Testing Population Designs," in *4th Metaheuristics International Conference Proceedings MIC'2001,* pp. 713–717.

Gutin, G. and A. P. Punnen (2002) *The Traveling Salesman Problem and its Variations,* Kluwer Academic Publishers, Boston.

Hamiez, J.P. and J.K. Hao (2001) "Scatter Search for Graph Coloring," *LNCS series,* Springer, forthcoming.

Hansen, P., N. Mladenovi´c and D. Pérez-Brito (2001), "Variable Neighborhood Decomposition Search," *Journal of Heuristics,* vol. 7, pp. 335-350.

Hart, J. P. and A. W. Shogan (1987) "Semi-greedy Heuristics: An Empirical Study," *Operations Research Letters,* vol. 6, pp.107-114.

Hill, R. R. and G. McIntyre. (2000) "A Methodology for Robust, Multi-Scenario Optimization," *Phalanx,* vol. 33, no. 3.

Holland, J. H. (1975) *Adaptation in Natural and Artificial Systems,* University of Michigan Press, Ann Arbor, MI.

Johnson, D.S. and M.A. Trick (1996) *Cliques, Coloring, and Satisfiability: 2nd DIMACS Implementation Challenge, 1993,* DIMACS Series in Discrete Mathematics and Theoretical Computer Science, vol. 26, American Mathematical Society.

Kelly, J., B. Rangaswamy and J. Xu (1996) "A Scatter Search-Based Learning Algorithm for Neural Network Training," *Journal of Heuristics,* vol. 2, pp. 129-146.

Kuo, C.C., F. Glover and K.S. Dhir (1993) "Analyzing and Modeling the Maximum Diversity Problem by Zero-One Programming," *Decision Sciences,* vol. 24, pp. 1171-1185.

Laguna, M., J. W. Barnes and F. Glover (1993) "Intelligent Scheduling with Tabu Search: An Application to Jobs With Linear Delay Penalties and Sequence-Dependent Setup Costs and Times," *Journal of Applied Intelligence,* vol. 3, pp. 159-172.

Laguna, M. and F. Glover (1993) "Integrating Target Analysis and Tabu Search for Improved Scheduling Systems," *Expert Systems with Applications,* vol. 6, pp. 287-297.

Laguna, M., P. Lino, A. Pérez, S. Quintanilla and V. Valls (2000) "Minimizing Weighted Tardiness of Jobs with Stochastic Interruptions in Parallel Machines," *European Journal of Operational Research*, vol. 127, no. 2, pp. 444-457.

Laguna, M. and R. Martí (1999) "GRASP and Path Relinking for 2-Layer Straight Line Crossing Minimization," *INFORMS Journal on Computing*, vol. 11, no. 1, pp. 44-52.

Laguna, M. and R. Martí (2000) "Experimental Testing of Advanced Scatter Search Designs for Global Optimization of Multimodal Functions," Technical Report, University of Colorado at Boulder, http://leeds.colorado.edu/faculty/laguna/articles/advss.html.

Laguna, M. and R. Martí (2001a) "A GRASP for Coloring Sparse Graphs," *Computational Optimization and Applications*, vol. 19, pp. 165-178.

Laguna, M. and R. Martí (2001b) "Neural Network Prediction in a System for Optimizing Simulations," *IIE Transactions*, forthcoming.

Laguna, M. and R. Martí (2002) "The OptQuest Callable Library," *Optimization Software Class Libraries*, S. Voss and D. L. Woodruff (Eds.), Kluwer Academic Publishers, Boston, pp. 193-218.

Laguna, M., R. Martí and V. Campos (1999) "Intensification and Diversification with Elite Tabu Search Solutions for the Linear Ordering Problem," *Computers and Operations Research*, vol. 26, pp. 1217-1230.

Lawler, Lenstra, Rinnoy Kan and Shmoys (1985) *The Traveling Salesman Problem: A Guided Tour of Combinatorial Optimization*, John Wiley and Sons.

Løkketangen, A. and F. Glover (1995) "Tabu Search for Zero/One Mixed Integer Programming with Advanced Level Strategies and Learning," *International Journal of Operations and Quantitative Management*, vol. 1, no. 2, pp 89-109.

Løkketangen, A. and F. Glover (1996) "Probabilistic Move Selection in Tabu Search for 0/1 Mixed Integer Programming Problems," in *Metaheuristics: Theory and Applications*, I.H. Osman and J.P. Kelly (Eds.), Kluwer Academic Publishers, pp. 467-488.

Løkketangen, A. and F. Glover (1998) "Solving Zero-One Mixed Integer Programming Problems Using Tabu Search," *European Journal of Operational Research*, vol. 106, pp 624-658.

Løkketangen, A. and F. Glover (1999) "Candidate List and Exploration Strategies for Solving 0/1 MIP Problems using a Pivot Neighborhood," in *Meta-Heuristics: Advances and Trends in Local Search Paradigms for Optimization*, S. Voß, S. Martello, I.H. Osman and C. Roucairol (Eds.), Kluwer Academic Publishers, pp 141–155.

Løkketangen, A. and D. Woodruff. (2000) "Integrating Pivot Based Search with Branch and Bound for Binary MIP's," *Control and Cybernetics*, vol. 29, no. 3, pp 741–760.

LOLIB (1997) http://www.iwr.uni-heidelberg.de/groups/comopt/software/LOLIB/index.html

Lourenço, H. R., J. P. Paixão and R. Portugal (2001) "Multiobjective Metaheuristics for the Bus Driver Scheduling Problem," *Transportation Science*, vol. 35, no. 3, pp. 331–343.

Martello, S. and P. Toth (1989) *Knapsack Problems*, John Wiley and Sons, Ltd., New York.

Martí, R. (1998) "A Tabu Search Algorithm for The Bipartite Drawing Problem," *European Journal of Operational Research*, vol. 106, pp. 558-569.

Martí, R. and M. Laguna (1997) "Heuristics and Metaheuristics for 2-Layer Straight Line Crossing Minimization," *Discrete and Applied Mathematics*, forthcoming.

Martí, R., M. Laguna and V. Campos (2002) "Scatter Search vs. Genetic Algorithms: An Experimental Evaluation with Permutation Problems," in *Adaptive Memory and Evolution: Tabu Search and Scatter Search*, C. Rego and B. Alidaee (Eds.), Kluwer Academic Publishers, forthcoming.

Martí, R., M. Laguna and H. Lourenço (2000) "Assigning Proctors to Exams with Scatter Search," in *Computing Tools for Modeling, Optimization and Simulation: Interfaces in*

Computer Science and Operations Research, M. Laguna and J. L. González-Velarde (Eds.), Kluwer Academic Publishers, Boston, pp. 215-227.

Martí, R., M. Laguna, F. Glover and V. Campos (2001) "Reducing the Bandwidth of a Sparse Matrix with Tabu Search," *European Journal of Operational Research*, vol. 135, pp. 450-459.

Mazzini, R. (1998) "Estudo de Meta-Heurísticas Populacionais para a Programação de Máquinas Paralelas com Tempos de Preparação Dependentes da Sequência e Datas de Entrega", Ph.D. Dissertation, Faculdade de Engenharia Elétrica e de Computação, Universidade Estadual de Campinas.

Michalewicz, Z. and T. D. Logan (1994) "Evolutionary Operators for Continuous Convex Parameter Spaces," in *Proceedings of the 3rd Annual Conference on Evolutionary Programming*, A. V. Sebald and L. J. Fogel (Eds.), World Scientific Publishing, River Edge, NJ, pp. 84-97.

Michalewicz, Z. (1996) *Genetic Algorithms + Data Structures = Evolution Programs*, 3rd edition, Springer-Verlag, Berlin.

Morgenstern, C.A. (1996) "Distributed Coloration Neighborhood Search," in *Cliques, Coloring, and Satisfiability: 2nd DIMACS Implementation Challenge, 1993*, Johnson and Trick (Eds.), DIMACS Series in Discrete Mathematics and Theoretical Computer Science, vol. 26, pp. 335-357.

Nelder, J. A. and R. Mead (1965) "A Simplex Method for Function Minimization," *Computer Journal*, vol. 7, p. 308.

Nowicki, E. and C. Smutnicki (1996) "A Fast Tabu Search Algorithm for the Job-Shop Problem," *Management Science*, vol. 42, no. 6, pp. 797-813.

Nowicki, E. and C. Smutnicki (2001a). "New Ideas in Tabu Search for Job-Shop Scheduling," Technical Report 50/2001.

Nowicki, E. and C. Smutnicki (2001b) "New Tools to Solve the Job-Shop Problem," Technical Report 51/2001.

Ochi, L. S., D. S. Vianna, L. M. A. Drummond and A. O. Victor (1998) "A Parallel Evolutionary Algorithm for the Vehicle Routing Problem with Heterogeneous Fleet," *Future Generation Computer Systems*, vol. 14, no. 5, pp. 285-292.

Oliva San Martin, C. (2000) "Scatter Search Pour le Problème du Voyageur de Commerce," M.S. Thesis, Université de Montréal.

Piñana, E., I. Plana, V. Campos and R. Martí (2001) "GRASP and Path Relinking for the Matrix Bandwidth Minimization," *European Journal of Operational Research*, forthcoming.

Prais, M. and C. C. Ribeiro (2000) "Reactive GRASP: An Application to a Matrix Decomposition Problem in TDMA Traffic Assignment," *INFORMS Journal on Computing*, vol. 12, pp. 164-176.

Press, W. H., S. A. Teukolsky, W. T. Vetterling and B. P. Flannery (1992) *Numerical Recipes: The Art of Scientific Computing*, Cambridge University Press (www.nr.com).

Rana, S. and D. Whitley (1997) "Bit Representations with a Twist," in *Proceedings of the 7th International Conference on Genetic Algorithms*, T. Baeck (Ed.), Morgan Kaufman, pp. 188-196.

Rego, C. and P. Leão (2000) "A Scatter Search Tutorial for Graph Based Permutation Problems," Technical Report, Hearin Center for Enterprise Science, School of Business Administration, University of Mississippi, USA.

Reinelt, G. (1985) *The Linear Ordering Problem: Algorithms and Applications*, Research and Exposition in Mathematics, vol. 8, H. H. Hofmann and R. Wille (Eds.), Heldermann Verlag Berlin.

Reinelt, G. (1994) *The Traveling Salesman: Computational Solutions for TSP applications*, Lecture Notes in Computer Science, Springer Verlag, Berlin.

Resende, M. G. C. and C. C. Ribeiro (2001) "Greedy Randomized Adaptive Search Procedures," in *State-of-the-Art Handbook in Metaheuristics*, F. Glover and G. Kochenberger (Eds.), Kluwer Academic Publishers, Boston, forthcoming.

Ribeiro, C.C., E. Uchoa and R.F. Werneck (2002) "A Hybrid GRASP with Perturbations for the Steiner Problem in Graphs," *INFORMS Journal on Computing*, forthcoming.

Rochat, Y. and É. D. Taillard (1995) "Probabilistic Diversification and Intensification in Local Search for Vehicle Routing," *Journal of Heuristics*, vol. 1, no. 1, pp. 147-167.

Roy, R. K. (1990) *A Primer on the Taguchi Method*, Van Nostrand Reinhold, New York.

Souza, M.C., C. Duhamel, and C.C. Ribeiro (2002) "A GRASP Heuristic Using a Path-Based Local Search for the Capacitated Minimum Spanning Tree Problem", Technical Report.

Ugray, Z., L. Lasdon, J. Plummer, F. Glover, J. Kelly and R. Martí (2002) "A Multistart Scatter Search Heuristic for Smooth NLP and MINLP Problems," in *Adaptive Memory and Evolution: Tabu Search and Scatter Search*, C. Rego and B. Alidaee (Eds.), Kluwer Academic Publishers, forthcoming.

Valls, V., M. Laguna, P. Lino, A. Pérez and S. Quintanilla (1998) "Project Scheduling with Stochastic Activity Interruptions," in *Project Scheduling: Recent Models, Algorithms and Applications,* Jan Weglarz (Ed.), Kluwer Academic Publishers, pp. 333-353.

Verhoeven, M.G.A. and E.H.L. Aarts (1995) "Parallel Local Search," *Journal of Heuristics*, vol. 1, pp. 43-65.

Voss, S., A. Martin and T. Koch (2001) "SteinLib Testdata Library," online document at http://elib.zib.de/steinlib/steinlib.html.

Woodruff, D. L. and E. Zemel (1993) "Hashing Vectors for Tabu Search," *Annals of Operations Research*, vol. 41, pp. 123-137.

Woodruff, D. L. (1996) "Chunking Applied to Reactive Tabu Search," in *Metaheuristics: Theory and Applications*, I.H. Osman and J.P. Kelly (Eds.), pp 555 – 570.

Woodruff, D. L. (1998) "Proposals for Chunking and Tabu Search," *European Journal of Operational Research*, vol. 106, pp. 585-598.

Woodruff, D. L. (2001) "General Purpose Metrics for Solution Variety," Technical Report, Graduate School of Management, UC Davis, Davis CA 95616, USA.

Xu, J., S. Chiu and F. Glover (2000) "Tabu Search and Evolutionary Scatter Search for 'Tree-Star' Network Problems, with Applications to Leased Line Network Design," in *Telecommunications Optimization: Heuristic and Adaptive Techniques*, D. W. Corne, M. J. Oates and G. D. Smith (Eds.), Wiley.

INDEX

T